Smart Innovation, Systems and Technologies

Volume 449

The Smart Innovation, Systems and Technologies book series encompasses the topics of knowledge, intelligence, innovation and sustainability. The aim of the series is to make available a platform for the publication of books on all aspects of single and multi-disciplinary research on these themes in order to make the latest results available in a readily-accessible form. Volumes on interdisciplinary research combining two or more of these areas is particularly sought.

The series covers systems and paradigms that employ knowledge and intelligence in a broad sense. Its scope is systems having embedded knowledge and intelligence, which may be applied to the solution of world problems in industry, the environment and the community. It also focusses on the knowledge-transfer methodologies and innovation strategies employed to make this happen effectively. The combination of intelligent systems tools and a broad range of applications introduces a need for a synergy of disciplines from science, technology, business and the humanities. The series will include conference proceedings, edited collections, monographs, handbooks, reference books, and other relevant types of book in areas of science and technology where smart systems and technologies can offer innovative solutions.

High quality content is an essential feature for all book proposals accepted for the series. It is expected that editors of all accepted volumes will ensure that contributions are subjected to an appropriate level of reviewing process and adhere to KES quality principles.

Indexed by SCOPUS, EI Compendex, INSPEC, WTI Frankfurt eG, zbMATH, Japanese Science and Technology Agency (JST), SCImago, DBLP.

All books published in the series are submitted for consideration in Web of Science.

Sergey V. Zykov

IT Crisisology Patterns and Practices

Smart Agility for Digital Future

Springer

Sergey V. Zykov
Graduate School of Business
HSE University
Moscow, Russia

ISSN 2190-3018 ISSN 2190-3026 (electronic)
Smart Innovation, Systems and Technologies
ISBN 978-981-95-1259-1 ISBN 978-981-95-1260-7 (eBook)
https://doi.org/10.1007/978-981-95-1260-7

This Springer imprint is published by the registered company Springer Nature Singapore Pte Ltd.
The registered company address is: 152 Beach Road, #21-01/04 Gateway East, Singapore 189721, Singapore

To God, my teachers, and my family

Foreword

When I am writing this Foreword, in March 2025, I would confidently say that we are living in an era of generalized crises. Media of any kind and in every language are constantly reporting about crises, being them geopolitical, climatic, social, sanitarian, educational, or financial. We are definitively living times where the word "crisis" seems overinflated. The original term "crisis" in English has been borrowed from Latin, which in turn was borrowed from Greek, meaning "discrimination, decision". The noun is derived from a verb meaning "distinguish, choose, decide". In English, crisis was first used in a medical context to then slowly become an equivalent for "decisive stage". In this period of time, it seems we are living decisive stages in many aspects of our individual and collective lives. In academia, crises themselves have been studied from diverse perspectives for centuries. Lately, the world "crisisology" has emerged as a discipline focusing on the study of crises. More recently, Sergey Zykov brought this ideas to the field of Information Technology and Software Engineering defining IT crisisology. In socio-technical systems, requirements are spanning from software/hardware to business and social. Understanding the social structure is essential in order to produce an informed system design. There are always three fundamental aspects that should be considered when approaching such kind of systems: technical, business, and human. The study of socio-technical systems is at the edge between software engineering, requirements engineering, and social sciences. IT crisisology is crisisology applied to the development of socio-technical systems and, more in general, to every information system. In a world with an increasing emphasis on resilience, IT crisisology emerged as the systematic study of crises in software production. Sergey Zykov over the years clearly emerged as the main authority in this field and as the creator of a fundamentally new discipline. Among the many books and articles that he dedicated to the topic, we can mention the "IT Crisisology Casebook" (2022) and the "IT Crisisology Models" (2024). In this new book, Sergey applies a systematic approach to digital production in critical circumstances. How to mitigate and manage crises by establishing a correct development process? This implies meeting quality standards and taking care of business requirements. Sergey, starting from intuitions and understanding inherited from more established engineering fields, leads the reader to the core of Crisisology in information systems, their lifecycle, the related development methodologies and the architectural aspects. "IT Crisisology Patterns and Practices: Smart Agility for Digital Future" should be a book present in the bookshelf of every Software Engineer, designer and requirements expert, as well as professionals in technical and project management. It can represent a valuable reading also for students at any level of education, from early undergraduates to advanced graduated studies in software engineering or related IT disciplines.

March 2025

Prof. Dr. Manuel Mazzara, Ph.D.
Full Professor in Software Engineering
Dean of Faculty of Computer Science and Engineering
Innopolis University
Innopolis, Russia

Acknowledgements

I would like to thank the colleagues of mine who significantly contributed to this book. They clarified my initially vague concepts and assisted in a number of processes including conceptualizing, translation, copyediting, diagramming etc. These are my current HSE colleagues Prof. Eduard A. Babkin and Dr. Boris V. Ulitin, and the students who did their master/Ph.D. theses and actively participated in the research seminars under my supervision. A few of their takeaways were essentially transformed and included into this book as patterns and practices for sustainable digital product development. They are Iskander Abdullakhanov, Ilya Chekushin, Adel Fattakhova, Yulia Milyakova, Indra Singh, Andrey Osipov, Daria Zakharova, Alexey Zheleznoy, and Maria Zolotoukhina.

I would like to thank the Springer Executive Editor Dr. Thomas Ditzinger, the Springer Senior Editor Mr. Aninda Bose, the Springer Senior Executive for Production Mr. Ashok Kumar, and the Springer Production Editor Mr. Karthik Raj Selvaraj, for their continuous availability and prompt assistance.

Last but not least, I would like to express my deep appreciation and sincere gratitude to the Editors-in-Chief of the Springer Series in *Smart Innovation, Systems and Technologies*, Prof. Lakhmi C. Jain and Prof. Robert J. Howlett, for their tireless efforts in supporting my ideas.

Introduction: Digital Development: People and Processes, Patterns and Practices…

This book is about applying a systematic approach to digital production in critical circumstances. Its key idea is mitigating and managing these circumstances by establishing and maintaining well-balanced development processes.

Given this context, "mitigating and managing" may really imply various activities such as predict, prevent, react to, adjust to, overcome, etc. To realize these processes, the digital products should not only meet a certain quality standards, i.e., be technologically suitable. The other mission-critical aspects of these prospective products must include both business requirements, which incorporate the company's structure and goals, and human factors. The interaction between these essential ingredients has been initially illustrated by the so-called Leavitt's diamond (see Fig. 1) [1]. However, a deeper dive into three kinds of mission-critical crisis factors, namely technical, business, and human, requires a more detailed framework, currently known as IT Crisisology (ITC) [2].

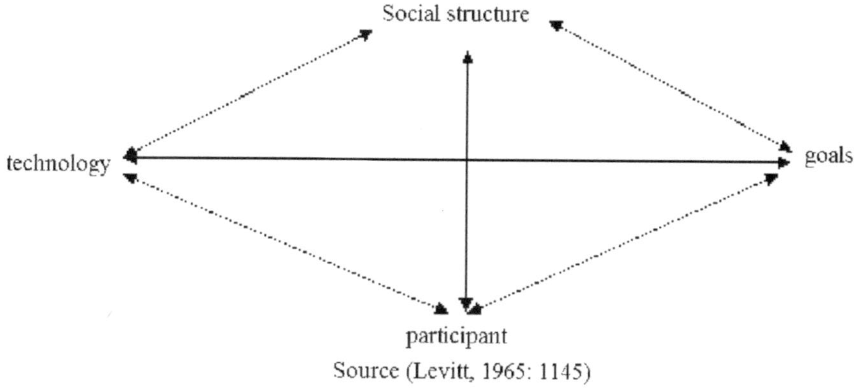

Fig. 1 Leavitt's diamond

Evidently, these three groups of factors are not only complex but also often contradictory. A common example would be operational performance and system security: fostering security hinders performance and vice versa. Thoughtful tradeoff management is required to ensure that these complex factors are actually effectively balanced. This kind of management should include both process improvement based on mathematically rigorous optimization methods and human-centered agile practices. The IT Crisisology framework prescribes a toolkit of useful techniques to adjust both. To illustrate our story, let us refer to a classic example of the Eiffel Tour in Paris, France [5]. One reason for that is the historical NATO 1968 conference decision to turn to architecture-centered methods in order to conquer the software development crisis which was already mission-critical even during those old times of the early digital production age [3]. Let us briefly recollect a few essential facts and figures.

Obviously, the tower is a world wonder, being not only a top landmark but also a technological masterpiece in concept, design, architecture, and construction. Located in the central part of Paris, this elegant wrought-iron building was intended to commemorate the French Revolution centennial. The author, chief designer and builder of this award-winning project selected out of 100 proposals, was Gustave Eiffel, a famous bridge construction engineer. An important note at this point is that in spite of many evident points of difference between the tower and a bridge, the project was a tremendous success for a number of reasons. First, Eiffel was an advanced construction engineer capable of blueprinting and guiding the entire production lifecycle. The lifecycle stages include conceptualizing, modeling, designing, prototyping, building, testing and maintaining this massive and complex object. Second, he expertly understood how the bridge construction principles and practices can efficiently transfer to support this entirely new and different object of architecture and art. Third, Gustav was a gifted manager capable of harnessing the human factors to stimulate the project progress.

As a result, the entire project lifecycle was completed in two years only (1887–1889). Moreover, it was very effective technically and financially. Currently, this tower exceeds 300 m (approx. 1000 feet) in height together with the TV antenna on top and weighs over 10,000 tons (approx. 20 million pounds). At that time, it was at least twice bigger than any other building, including the famous Great Pyramid of Giza and even St. Peter's Cathedral in Rome.

The reader could ask at this point, what was the secret of this cost-effective success? Upon careful mathematical modeling, Eiffel's idea was to gradually scale his open-lattice high-level blueprint down to the mid-size (four lattice-girder pier stands, two-level girders, etc.) and elementary combinability for the tower ensemble. Moreover, he suggested an efficient scale-up assembly process based on the properties of these elementary details, including over 18,000 of iron parts and around 2,500,000 of rivets. This scalable process allowed for surprisingly fast, cheap, and reliable construction of the entire object, both aesthetically beautiful and technically strong. Conversely, erecting St. Peter's Cathedral took over a century, and the Great

Pyramid of Giza required enormous slave labor and resulted in multiple incidents. Therefore, a brief answer to the above reader's question is: the "secret" is scalable and reliable patterns and practices based on strategic reuse of the models, principles, techniques, and elements such as rivets.

Another positive strategic reuse example of a less known design concept and pattern by a prominent architect related to Eiffel Tower was the Sheepford Road Bridge. This bridge was erected in 1887 by Phoenix Bridge Company located in Phoenixville, PA, USA, along the Yellow Breeches Creek in Pennsylvania. This design is famous for the so-called "Phoenix columns" support, later reused for many architectural objects including the famous Eiffel Tower [5]. Gustav Eiffel had credited the Phoenix Bridge Company engineer, Samuel Reeves, for this brilliant concept that he successfully reapplied in his Parisian masterpiece the construction of which started the same year and was extremely efficient in terms of time, cost, and labor.

This book is structured as follows. Chapter 1 provides a high-level pattern-based layout of digital product development. The essential ingredients of this overview are general principles, handy techniques, and best practices. Research is focused on the simplicity of the foundation and overall scalability, resulting in operational efficiency similar to the Eiffel Tower example. Based on the ITC approach, these advanced decision-making practices, tradeoff optimization techniques, and other methods provide the patterns for systematically addressing the business, technical, and human-related factors.

Chapter 2 discusses the methods and practices to enhance transaction management quality in large and therefore complex IT infrastructures. This kind of complexity is a possible crisis trigger and may be caused by either encumbering organizational layout, type heterogeneity, or a different reason. To support the implementation, we recommend applying the DevOps methods and techniques, such as continuous integration/delivery and automated testing. Optimization-based formal model serves as the transfer process foundation and customer survey validates its outcomes. Instantiating the general pattern of the IT Crisisology process for digital transfer management, the approach suggested includes a typical layout of rigorous models, flexible methods, and computer-aided instruments.

Chapter 3 presents a practically oriented case of oil-and-gas company digitalization based on process improvement for hydraulic fracturing. Such process improvement is unavoidable and mission-critical for addressing uncertain and rapidly changing businesses, featuring complex and diverse landscapes. Combined, such landscapes together are likely to form an IT crisis minefield. Similar to the previously mentioned cases, this dramatic digital change requires well-adjusted techniques and methods, which certainly may be based on the IT Crisisology framework. In the case of oil-and-gas business, requirements and constraints become cross-cutting dependencies that intensively protrude through the IT infrastructure landscape. This brings together several data/information levels, such as data acquisition and aggregation (including digitizing, filtering, normalizing, and cleaning); daily, midterm, and long-term management, strategic resource planning and business analysis, and top-level decision-making. To combine the human, technology, and business factors, the process improvement pattern was data-centered and featured a smart business-intelligent dashboard to visually present the high-level KPI dynamics to the key stakeholders and decision makers. This efficient tool assisted in business-focused smart data integration and on-the-fly scenario switching (based on visual hypercube slicing, dicing, pivoting, etc.), which resulted in critical business process improvement, better top-level decision-making and, finally, digital transformation success.

Chapter 4 addresses large-scale digitalization financial business transformation based on a set of cases including Raiffeisen among a few other world recognizable multinationals. In such cases, the common crisis causes often involve structural complexity and stiff competition. However, this story would be incomplete without mentioning the technical problems and some other issues. To harness these diverse issues, IT Crisisology prescribes tradeoff optimization for financial portfolio management. This instance of the framework includes such smart ingredients as sentiment analysis, large language model, and visual presentation. Validation of the resulting API-based pattern for the digital transfer by means of customer survey of the digital product provided evidence of better balance for such critical system quality attributes as performance, accuracy, elasticity, and reliability.

Chapter 5 elaborates on digital landscape transformation in complex service-oriented organizations. Along with complexity, the factors that these organizations have to compromise on include intensive multitasking, sharp rivalry, and environmental volatility. This given, complex corporate activities coupled with multifaceted business diversity become a potential crisis trigger. To prevent this misfortune, this chapter suggests flexible techniques and methods to improve the IT infrastructure based on an iterative feedback-oriented lifecycle. To fill this high-level pattern, the authors suggest a standardized process framework set that includes BPMN, UML/IDEF and ITIL/ITSM approaches, and the Draw.io as a CASE tool. This pattern was systematically used to digitally improve/optimize business processes in several IT servicing companies. The results showed dramatic time saving and much better customer satisfaction.

Chapter 6 suggests a digital pattern to integrate big data into a large-scale multi-format landscape. This landscape contains a set of warehouses and data management systems. Given this context, the data bulk, along with its rapid changeability and heterogeneity makes a likely and potentially dangerous crisis source. In terms of components i.e., architecturally, the flexible unification pattern suggested for this multifaceted data layout is based on the extract-transform-load (ETL) process and layer organization of the solution structure. To make the solution agile, the above pattern integrates metadata modeling and management. Zooming in, this approach features a microservice-based detailed design, subscription event management, and an API-type interaction to promote loosely coupling and crisis-responsive data management. The other essential aspect of the approach incorporates orchestration to efficiently manage multitype data. The solution is supported by a smart set of tools including Apache Airflow, Kubernetes, and ML-based recommender to allow interfacing with multiple social networks.

Chapter 7 introduces the jerky dynamics of digital transformation due to rapid infrastructural landscape upgrades. Agile approaches such as Kanban, Lean, Scrum, and SAFe are used as a positive pattern compared to digital crises characterized by ill-balanced dynamics. Before summarizing the pitfalls in the final chapter and concluding remarks, the Waterfall approach is identified as a commonly encountered antipattern of an extremely inflexible IT-intensive process and, therefore, a crisis trigger. Along with the IT Crisisology, DevOps is recommended as the common framework for the patterns and practices suggested. In addition to the above-mentioned models and methods, the patterns designed incorporate a set of tools to promote such essential digitalization processes as communication/collaboration, configuration management, continuous development/integration, testing, and monitoring. Emerging trends in these changeable hi-tech landscapes, such as blockchain, quantum computing, privacy/security management, and smart automation, are discussed.

The conclusion recaps the research outcomes and discusses the emerging trends in sustainable digital production.

We wish you a satisfying trip through the book and hope it provides a powerful toolkit in developing digital solutions and overcoming any crisis-related obstacles by handy patterns and useful practices.

References

1. Laudon, K. C., & Laudon, J. P. (2016). Management Information Systems: Managing the Digital Firm. Pearson
2. Zykov S. V. (2021) IT Crisisology: Smart Crisis Management in Software Engineering. Models, Methods, Patterns, Practices, Case Studies Issue 210. Singapore: Springer, 2021. ISBN 978-981-33-4434-1, https://doi.org/10.1007/978-981-33-4435-8
3. Naur, P., Randell, B. (ed.): Software Engineering: Report on a Conference sponsored by the NATO Science Committee, Garmisch, Germany, 7th to 11th October 1968, Brussels, Scientific Affairs Division, NATO, 231 pp., January 1969
4. Randell, B.: Software engineering in 1968. In: Proceedings of the 4th International Conference on Software Engineering, pp. 1–10, Munich (1979)
5. Eiffel Tower: History, Height, & Facts (2024) Britannica Online. Retrieved November 25, 2024 from https://www.britannica.com/topic/Eiffel-Tower-Paris-France

About This Book

This book paves way for pattern-based practices of digital product development, particularly adjusted for crisis conditions. It uses the foundation of the IT Crisisology, which includes monitoring, predication, prevention, and resilient management of these unfavorable events.

The recent ways of outlining the lifecycles for digital production often incorporate multifaceted approaches and complex landscapes. These include devastating business constraints, volatile technical requirements, and unstable human factors, each of which being a potential crisis trigger and a likely reason to slow down and maybe even impede and complicate the digital transformation. Adequate management of these complex issues at a moderate cost requires a process-based strategy. Based on the framework of the IT Crisisology models and methods, this book specifies its "last mile", i.e., a set of essential patterns and helpful techniques that provide business agility and ensure digitalization success even in critical conditions.

This "Swiss-army-knife" toolkit of crisis-resilient patterns and practices encompasses a thoughtfully selected blend incorporating infrastructure-based libraries, process optimization approaches, and agile communication methods This book offers system building principles and software architecture examples that will help the developers to justify their own reasons in order to avoid potential shortcomings in digital products and advocate intelligent construction. As this IT Crisisology-based approach successfully escalates to support bulky and complex systems, this volume is aimed to provide help in crisis-responsive techniques, concentrating on their utilizable production that promotes smart and sustainable digitalization.

To intelligently apply the patterns and practices in uncertain environments and changeable conditions, this book introduces a set of realistic and real-world case studies. Their backgrounds range from classical standards, such as ITIL and BPMN notations for business process improvement and ascend to human and architecture-centered modern approaches, such as subject-oriented BPM, microservices, DevOps, and their state-of-the-art implementations.

Upon meditating on these cases and identifying their most probable scenario-based extensions, the developers can further custom-tailor, adjust, and apply the relevant patterns and practices to match their mission-critical requirements.

We sincerely hope that this concise book provides a handy toolkit of useful principles and strategies to the digital product participants to help them confidently navigate in the rough seas of business digitalization.

Contents

Acronyms

A3C	Asynchronous Advantage Actor-Critic
ABPMP	Association of Business Process Management Professionals
ACDM	Architecture Centric Design Method
ACDM/ATAM	Architecture Centric Development Method
ACID	Atomicity, Consistency, Isolation, Durability
ACM	Association for Computing Machinery
ACO	Ant Colony Optimization
AHP	Analytic Hierarchy Process
AI	Artificial Intelligence
AICSM	Artificial Intelligence Cyber Security Management
AIMA	Automated Iterative Messaging Algorithm
AL	Artificial Intelligence
AMQP	Advanced Message Queuing Protocol
ANN	Artificial Neural Networks
API	Application Programming Interface
BASE	Basically Available, Soft state, Eventually consistent
BERT	Bidirectional Encoder Representations from Transformers
BI	Business Intelligence
BP	Business Process
BPM	Business Process Management
BPMN	Business Process Model and Notation
CAM	Categorical Abstract Machine
CaML	Categorical Meta Language
CAP	Consistency, Availability, Partition tolerance
CBOK	Common Body of Knowledge
CI/CD	Continuous Integration/Continuous Deployment
CMMI	Capability Maturity Model Integration
COBIT	Control Objectives for Information and Related Technologies
COVID	Coronavirus Disease
CRUD	Create, Read, Update, Delete
CVE	Common Vulnerabilities and Exposures

DE	Data Engineering
DFD	Data Flow Diagram
DNA	Deoxyribonucleic acid
DNS	Domain Name System
DPD	Digital Product Development
DQ	Data Quality
DQNs	Deep Q-Networks
DTP	Distributed Transaction Processing
DWH	Data Warehouse
E5	Enterprise 5
ELK	Elasticsearch, Logstash, Kibana
EO	Elegant Objects
EOR	End of Run
ERA	Evidential Reasoning Approach
ERP	Enterprise Resource Planning
ES	Evolution Strategy
ESB	Enterprise Service Bus
ETL	Extract, Transform, Load
GA	Genetic Algorithms
GDPR	General Data Protection Regulation
GP	Genetic Programming
GPT	Generative Pre-trained Transformer
HATEOAS	Hypermedia as the Engine of Application State
HTTP	Hypertext Transfer Protocol
IaC	Infrastructure as Code
IAS	Iterative Averaging Smoothing
IDEF0	Integration Definition for Function Modelling
IDEF3	Integration Definition for Process Description
IDEs	Integrated Development Environments
IoT	Internet of Things
IP	Internet Protocol
IS	Information System
ITC	IT Crisisology
ITC(F)	IT Crisisology (Framework)
ITIL	Information Technology Infrastructure Library
ITIL/ITSM	Information Technology Infrastructure Library/IT Service Management
ITSM	IT Service Management
JTA	Java Transaction API
JVM	Java Virtual Machine
KPI	Key Performance Indicator
LDA	Latent Dirichlet Allocation
LLT	Low-Latency Trading
MC	Microcontroller
MCDA	Multiple-Criteria Decision Analysis

MES	Manufacturing Execution System
ML	Machine Learning
MOF	Model-Driven Architecture
MSIL	Microsoft Intermediate Language
NLP	Natural Language Processing
OOP	Object-Oriented Programming
ORM	Object-Relational Mapping
PBTL	Population-based Translational Learning
PCA	Principal Component Analysis
PDCA	Plan-Do-Check-Act
PM	Project Management
PoC	Proof of Concept
PSO	Particle Swarm Optimization
QA	Quality Attribute
RDBMS	Relational Database Management System
REST	Representational State Transfer
REST or gRPC	Representational State Transfer or gRPC
RL	Reinforcement Learning
RLHF	Reinforcement Learning from Human Feedback
RPC	Remote Procedure Call
RSSE	Requirements Specification and Software Engineering
SA	Software Architecture
SCADA	Supervisory Control and Data Acquisition
SDLC	Software Development Lifecycle
SECD	Object-Based Abstract Machines
SEO	Self-Organization
SFT	Supervised Fine-Tuning
SOA	Service-Oriented Architecture
SQL	Structured Query Language
SSL/TLS	Secure Sockets Layer/Transport Layer Security
STOMP	Simple (or Streaming) Text Orientated Messaging Protocol
SWOT	Strengths, Weaknesses, Opportunities, Threats
T5	Text-to-Text Transfer Transformer
TDD	Test-Driven Development
UIF	User Interface
UML	Unified Modeling Language
UML/IDEF	Unified Modeling Language/Integration Definition
VIM	Vulnerability Information Management
VIP	Virtual IP
VLDB	Very Large Database
VTB	Virtual Training Base
XML	Extensible Markup Language

Chapter 1
Digitalization: Crisis Trigger or Remedy?

Abstract This chapter provides a digitalization overview in terms of the pattern-based approach. As such, it discusses common digitalization principles, techniques, and practices. The framework used is IT Crisisology from our earlier books by Springer, which incorporates business, technical and human-related factors. The methods discussed include decision-making techniques, trade-off optimization, and a few other approaches. These often simple yet efficient methods focus on crisis management in digital transfer. Further, the chapter outlines the pattern-and-practice framework to be used and analyzes its strengths and weaknesses. To prove the concept, the suggested principles and practices are verified by business-related case studies.

1.1 The Concept of Methodology in the Context of the Development of Information Systems

Methodology is a set of methods, techniques and sequential actions aimed at achieving a set of goals. In the context of information systems (IS) lifecycle planning, the methodology includes strategies and approaches that ensure the successful creation, implementation, operation and support of IS at all stages of its lifecycle.

The technique may include:

1. Requirements development—methods for collecting and analyzing business needs to determine the functionality of a system
2. Design—approaches to creating system architecture, data models, and interfaces
3. Development and testing is a sequence of actions for writing code and checking its validity
4. Implementation and operation—strategies for change management, system deployment and support
5. Upgrade and disposal—methods for updating the system in a timely manner and completing its lifecycle.

S. V. Zykov, *IT Crisisology Patterns and Practices*, Smart Innovation, Systems and Technologies 449, https://doi.org/10.1007/978-981-95-1260-7_1

The methodology helps to structure the process, reduce risks and improve quality at all stages of the lifecycle.

1.2 The Concept of an Information System and Its Characteristics

The concept of an information system includes a wide range of components and interactions that are vital for the functioning of an organization. For IT systems lifecycle management, it is important to understand how these systems are designed and implemented to meet changing technology. An information system is generally defined as an integrated set of components designed to collect, process and disseminate data. These systems typically include hardware, software, data, and human resources.

IS can be described through the following characteristics:

Interactivity

Information systems are inherently interactive, allowing users to input, retrieve, and manipulate data in real-time. This interactivity is critical to operational efficiency and responsiveness, allowing decision-makers to quickly access updated information. Studies emphasize that the design of the user interface significantly affects the effectiveness of interaction, affecting the overall usability of the system and user satisfaction [1].

Scalability

The ability of an information system to scale to meet the needs of an organization is vital. Scalability ensures that the system can handle the growth in the number of users, data volume, and transaction volume without sacrificing performance. Various architectural aspects that increase IS scalability, such as cloud computing technologies, allow systems to dynamically expand resources as demand increases [2].

Flexibility

The flexibility of information systems is related to their ability to adapt to new business processes and regulatory requirements. This adaptability is becoming increasingly important as organizations face rapid technological change and competitive pressures. The modular design of the system can support flexibility, allowing components to be added, removed, or modified with minimal impact on the system as a whole [3].

Integration

Integration involves the seamless connection of various information systems, both inside and outside the organization. Effective integration contributes to the unification of data flows and decision-making processes. System integration can lead to improved organizational consistency and process efficiency, emphasizing the role of middleware and enterprise application integration technologies [4].

Security

Security is essential to protect information systems from unauthorized access, data breaches, and other cyber threats. The security of an information system must ensure the confidentiality, integrity, and availability of data. The implementation of fundamental security principles, such as minimization of privileges and separation of duties, is crucial for the creation of reliable systems [5].

Data Management

Reliable data management is a hallmark of effective information systems. Proper data management ensures the accuracy, completeness, and reliability of the data stored and processed. Data quality goes beyond mere correctness and encompasses aspects such as completeness, consistency, timeliness and relevance. Data consumers, including decision-makers and analysts, prioritize these additional dimensions based on their specific needs and context. For example, timely data may be more important in certain operational scenarios, while consistency and completeness may be paramount for analytical tasks [6].

Taking into account these characteristics in the design and operation of information systems is essential to manage the lifecycle of information systems in a changing technological landscape. Each of these aspects contributes to the efficiency and agility of the system in dynamic environments, ensuring its long-term relevance and value.

1.3 Current State and Problems of the IS Lifecycle

Organizations rely heavily on IS to support their business processes, decision-making, and competitive strategies. Today's IS landscape reflects the convergence of various technologies, including cloud computing, big data analytics, the Internet of Things (IoT), and artificial intelligence (AI). These technologies have changed the way organizations collect, process, and use information, driving innovation and efficiency across all industries.

The widespread use of cloud computing in modern information systems allows organizations to dynamically scale resources, reduce infrastructure costs and increase operational flexibility. Similarly, advances in big data analytics have enabled organizations to extract valuable insights from large amounts of data, enabling data-driven decision-making and personalization of the customer experience.

Despite the transformative potential of IS, organizations face a number of challenges in managing the IS lifecycle:

Technological obsolescence

The rapid pace of technological change often leads to the obsolescence of IS components and architectures. Legacy systems become incompatible with new technologies, which makes interoperability and integration difficult.

Security Hazards

The increasing digitalization of data and processes exposes information security to various security threats, including cyberattacks, data breaches, and insider threats. Many studies highlight the growing complexity of cyber threats and the need for robust security measures throughout the IS lifecycle, including encryption, access control, and threat detection systems.

Complexity of data management

The exponential growth of data poses challenges in terms of storage, management, and analysis. Organizations are trying to harness the value of their data while ensuring its quality, integrity, and confidentiality. The effectiveness of data management systems and data management practices to address these challenges is of high importance.

Flexibility and adaptability

Traditional IS development methodologies, such as Waterfall, often lack the flexibility needed to respond quickly to changing business requirements and technological advances. The Agile and DevOps approaches have emerged as alternatives, with a focus on iterative development, collaboration, and rapid deployment.

Legacy systems

Many organizations are faced with outdated, poorly documented and difficult to maintain information systems. Modernization or replacement of these systems is associated with significant difficulties due to time costs and failures in operation. Many studies advocate a strategic approach to modernizing legacy systems, taking into account factors such as business value, technical feasibility, and organizational readiness.

Addressing these challenges requires a multifaceted approach to IS management, including strategic planning, effective governance, robust security measures, and agile development methodologies. Moreover, organizations should develop a culture of innovation, collaboration, and continuous improvement, in order to effectively manage the complexities of the IS lifecycle.

1.4 IS Lifecycle Models

Information systems (IS) lifecycle models provide a structured framework for the systematic management of the development, deployment, and maintenance of information systems. These models ensure that systems effectively meet business requirements and adapt to changing technologies and business environments. The following excerpt examines the various common models of the IS lifecycle, highlighting their features, areas of application and the results of the latest research.

Waterfall model

The Waterfall model, also known as the cascade model, is a linear and consistent approach to software development. It was first introduced by Winston W. Royce, in 1970, as a structured systems development methodology. The model is called "cascade" because progress is visualized as a cascade of individual stages, each of which depends on the completion of the previous one (Fig. 1.1).

The Waterfall model consists of the following stages, each of which has clearly defined goals and outcomes:

1. Requirements Analysis: This is the foundational stage where project requirements are gathered from stakeholders. Requirements are documented in detail to guide future phases.
2. System Design: Based on the requirements specification, the architecture and design of the system are developed, describing hardware and software components, data models, and user interfaces. During the design phase, system design documents are developed to guide implementation.

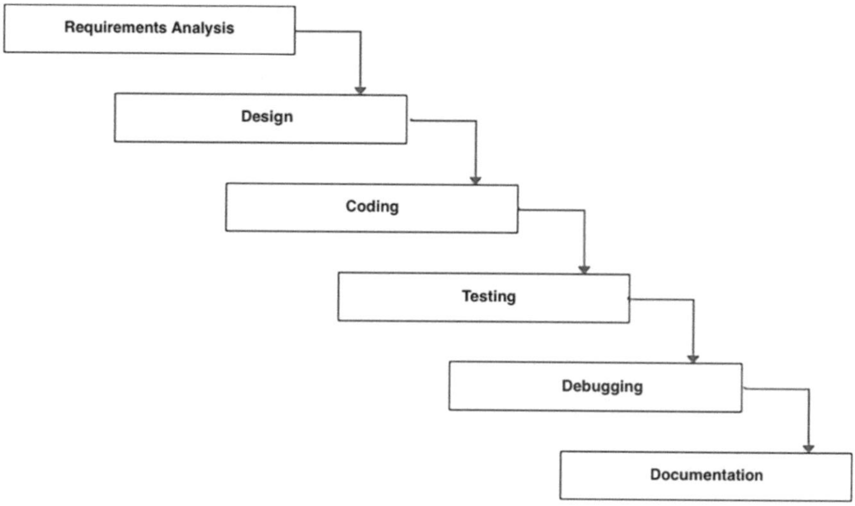

Fig. 1.1 Waterfall model diagram

3. Implementation (coding): This stage involves translating the design of the system into code through programming. Individual software components are coded and then integrated according to the system architecture.
4. Testing: After implementation, the system undergoes thorough testing to identify and eliminate errors. Testing ensures that the system meets the original requirements and works according to specifications.
5. Deployment: After successful testing, the system is deployed to the target environment, where it is made available to users.
6. Maintenance: In this phase, maintenance of the deployed system is carried out to address issues such as bugs, security vulnerabilities, and feature requests.

Advantages and disadvantages

The strengths of the Waterfall model lie in its structured and disciplined approach, which works well for projects with stable and well-understood requirements. It is easier to manage due to its predictability and clearly defined stages.

The most significant drawback of the model is the rigidity in adapting to changing requirements. Changing requirements, after development and implementation, often leads to significant reworking and an increase in costs. Such rigidity makes it less suitable for projects in which requirements are expected to change.

Despite its limitations, the Waterfall model remains relevant in regulated industries and projects with clearly defined requirements. Recent changes include feedback loops that allow for iterative refinement within stages and emphasize early validation of requirements. These changes aim to address criticisms by providing some flexibility while maintaining a general structured approach.

While the Waterfall model is not as adaptable as today's iterative and agile models, it is still useful for projects that require strict compliance and where the requirements are unlikely to change significantly.

V-model

The V-model (Verification and Validation Model) is a software development methodology that extends the Waterfall model by emphasizing parallel testing activities corresponding to each stage of development. It is named after the V-shaped structure, which illustrates the relationships between the development phases on the one hand and the corresponding testing phases on the other.

The V-model is divided into two main streams: development (to the left of the "V") and testing (to the right of the "V"). The steps are as follows:

Development stages:

1. Requirements analysis

Business requirements and system requirements are collected from stakeholders and analyzed to create a detailed requirements document.

Eligibility criteria are established to determine whether the final system meets the needs of stakeholders.

2. System design

High-level design of the entire system, breaking down requirements into logical subsystems or components.

3. Architectural Design

The technical architecture and overall structure of the system are defined, including data flows, external interfaces, and key components.

4. Module Design

The design of individual modules is described in detail, which provides programmers with recommendations for implementation.

5. Implementation

The system is coded, each module is implemented on the basis of the module's design documentation.

Stages of testing

1. Unit Testing

Individual modules or units are tested to verify their internal logic and functionality, according to the module design.

2. Integration testing

Integration tests verify the proper interaction between modules based on the architecture and design of the system.

3. System Testing

System tests check the integrated system against the system design to ensure that all features and requirements are met.

4. User Acceptance Testing (UAT)

The final acceptance tests confirm that the system satisfies the business requirements specified in the requirements analysis.

Advantages and disadvantages

The V-model has a number of advantages due to its structured nature and parallel verification/validation steps:

- Early detection of defects: Testing begins with the requirements phase, which allows for the timely detection and elimination of defects.
- Clear documentation: At each stage, documentation is created that guides the subsequent stages, ensuring traceability.

Despite its strengths, the V-shaped model still remains inflexible in a rapidly changing environment and requires well-defined prerequisites. Its lack of adaptability makes it less suitable for projects with changing or uncertain requirements, especially in a rapidly changing technology landscape.

Recent developments

Recent changes to the V-model include iterative feedback loops, which allow for limited changes at different stages, increasing flexibility. These changes align the V-model more closely with the agile and iterative principles, while maintaining a disciplined approach.

In conclusion, while the V-model may not be suitable for every type of project, it remains a valuable approach for projects where compliance, early detection of defects, and clear requirements are required.

The iterative model is an approach to software development characterized by incremental development and improvement through iterative cycles. Each cycle, or iteration, is aimed at improving the functions of the system, which makes the model suitable for projects whose requirements are changing or are initially ambiguous (Fig. 1.2).

Stages

1. Requirements Definition and Planning
   - At each iteration, requirements are defined and prioritized. These can be new features or improvements to existing ones.

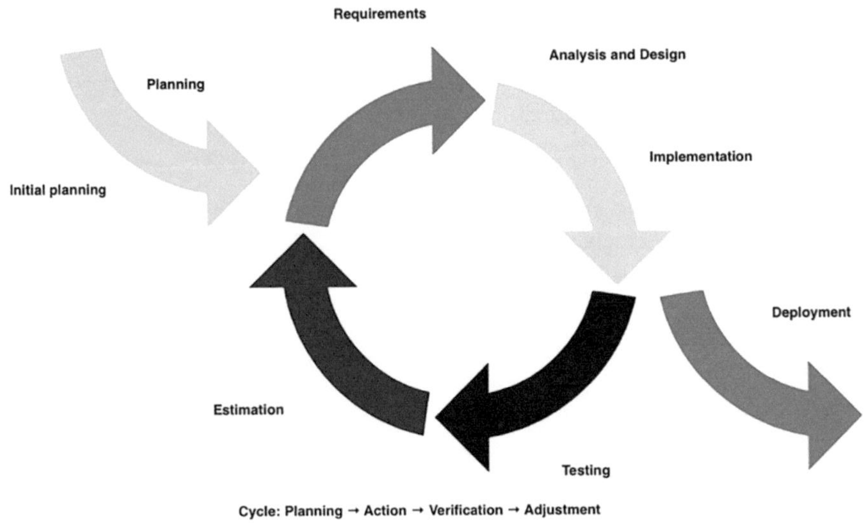

Fig. 1.2 Iterative model

  - Planning involves deciding which requirements will be met in the current iteration based on their importance and technical feasibility.

2. Design and development
   - To meet the requirements of the current iteration, a preliminary design is created based on the existing architecture and adds new modules.
   - The design is translated into code, and the developers implement certain features.

3. Testing and evaluation
   - Testing is done to verify that newly developed features work as expected and integrate well with the existing system.
   - Where possible, feedback from users is sought to identify areas for improvement.

4. Review and refinement
   - At the end of each iteration, the progress of the project is discussed with stakeholders.
   - Lessons learned from the current cycle are then used to refine plans for the future.

Advantages

Flexibility: The iterative model allows for continuous improvement of requirements and functions, providing flexibility in project management.

Risk Reduction: Early identification of problems reduces overall risk by gradually resolving problems.

Restrictions

However, iterative development can lead to scaling up if project management is not rigorous in prioritizing and finalizing requirements.

The iterative model is often used in projects where the initial requirements are uncertain or are likely to change. Modern methodologies, such as Agile, have adapted their principles to practices such as Scrum and Kanban, which emphasize frequent feedback and continuous improvement. Rational Unified Process (RUP) and Dynamical Systems Design Method (DSDM) also use iterative development, although each has its own characteristics and structure.

Recent developments

Recent literature points to the growing integration of iterative development with continuous delivery and integration practices. This allows for faster feature deployment and early user validation. Modern DevOps practices further enhance iterative models by enabling automation and rapid feedback mechanisms.

Spiral model

The spiral model is an agile approach to software development based on iterative risk management.

Although the spiral model is similar to the iterative model, there is a difference between the two. Spiral and iterative models are software development methodologies that emphasize operational improvement and evolution, but they differ in their specific development approaches. The spiral model is a more structured and risk-oriented methodology, often suitable for high-stakes projects (Fig. 1.3).

Stages and structure

Each cycle i.e., spiral of the model consists of four main stages:

1. Setting goals and identifying alternatives
   - Goals for the current iteration are determined, taking into account user requirements, project constraints, and risk factors.
   - Alternatives for achieving goals have been studied.
2. Risk assessment and mitigation

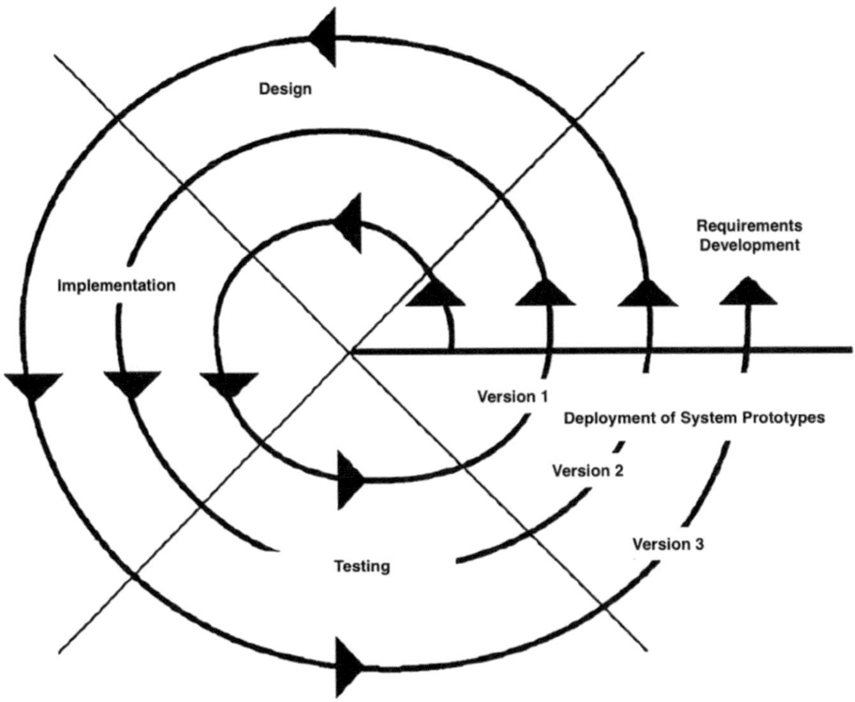

Fig. 1.3 Spiral model

- The potential risks associated with each alternative are evaluated and strategies are developed to mitigate them through prototyping, modeling, or other methods.

3. Development and verification
   - Prototypes or system modules are developed based on the results of a risk assessment.
   - Results are reviewed and validated to ensure that they meet requirements and objectives.
4. Analysis and planning
   - The results of the current iteration are discussed with stakeholders.
   - Planning for the next cycle is adjusted based on feedback and updated risks.

Benefits and constraints

- Risk mitigation: The spiral model remains relevant due to its structured risk identification and management strategies, which lead to early detection and mitigation.
- Iterative refinement: It provides flexibility through repetitive development cycles, allowing adjustments to be made based on user feedback or changing design requirements.

However, the use of the model in specialized risk analysis and management processes makes it more suitable for large projects with sufficient resources and budgets.

Application and adaptation

Recent implementations have successfully combined the spiral model with agile methodologies to manage complex projects at risk. For example, the Agile Incremental Commitment Spiral Model (AICSM) integrates agile practices while maintaining the risk orientation inherent in the Agile model. Other organizations are using this model to implement large-scale IT infrastructure projects in banking, defense, and healthcare.

1.5 Development Methodologies: Technological Landscape

Software systems development methodologies include structured approaches that define the processes of creating, deploying, and maintaining software. These frameworks offer specific methods, practices, and principles designed to manage the complexities of software projects by providing a systematic means for planning, structuring, and monitoring development activities. There are many benefits to implementing a software development methodology, including an improved structure and organization that clarifies roles and task sequencing; improving quality and reliability through standardized testing and validation steps, and improving efficiency and productivity by identifying best practices. In addition, these methodologies promote improved communication and collaboration between development teams and stakeholders, include risk management strategies to address potential issues at an early stage, and provide mechanisms for tracking and controlling projects, ensuring that projects are on track, on budget, and on schedule.

1.6 Existing Development Methodologies

Popular methodologies such as Agile, Scrum, Kanban, Waterfall, and DevOps cater to different project needs, but generally emphasize adaptability, efficient workflow management, and stakeholder satisfaction. These methodologies are indispensable to ensure that software development projects are completed on time, within budget, and achieve high quality, while remaining flexible enough to adapt to changing requirements and technologies.

Let us take a closer look at the most popular methodologies.

Agile

Agile is a project management method based on an iterative approach, with an emphasis on collaboration, flexibility, customer feedback, and getting tasks done quickly. Originally developed for software development, it is now applied in various fields to improve efficiency and adaptability to ever-changing customer requirements.

Basic principles and practices

Iterative development

Agile divides a project into small parts—sprints or iterations that last several weeks. In each sprint, the team goes through all stages of development, which allows you to regularly revise plans and clarify details.

Cooperation and feedback

Constant interaction with customers and other stakeholders is an important part of Agile. This allows the project to meet current user requirements and make changes as needed.

Flexibility and adaptability

The Agile methodology is designed to easily accept changes at any stage of the project, which makes it especially effective in a rapidly changing technological landscape.

Stages of the development cycle (Fig. 1.4).

Project planning

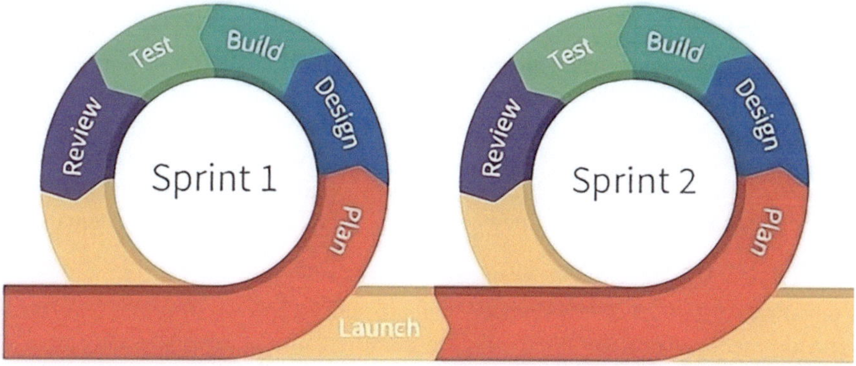

Fig. 1.4 Agile cycle

At this stage, the goals, scope, and feasibility of the project are determined. Requirements are collected and an initial roadmap is developed. A team is then formed that is responsible for various aspects of development, testing, and customer interaction.

Release planning

The project is divided into iterations or sprints. A prioritized task list is created and regularly updated throughout the project.

Sprint cycle

Each sprint lasts from 2 to 4 weeks. At the beginning of the sprint, the team selects tasks to complete, holds daily meetings to track progress and discuss problems, and at the end of the sprint, demonstrates the results and analyzes the process for improvement.

Release and evaluation

Completed features are integrated into the product, which is then tested to ensure its quality. Clients provide feedback, which can then influence the further development of the project.

Completion of the project

Once all tasks are completed, the final version of the product is released. The team reviews the project and develops plans for further support and maintenance.

Agile focuses on flexibility and adaptability, allowing the team to respond quickly to changes and adjust the project to meet customer needs.

In recent discussions on agile methodologies, several key ideas have emerged, in particular, regarding their adaptation to different organizational contexts and their integration with other methodologies. These insights reflect the agility of Agile and its relevance in the rapidly changing technology landscape, which is critical to information systems lifecycle planning.

1. Flexible customization in the development of information systems

Researchers emphasize the adaptability of agile methodologies to the various needs of organizations. Agile principles allow you to adapting them to the requirements of specific projects and organizational structures. Such individualization is especially valuable in the development of information systems, where technological and business environments are constantly evolving. Agile's ability to respond to change helps organizations keep abreast with new technologies and changing market demands.

2. Agile and DevOps integration

Integrating Agile with DevOps practices significantly improves information systems lifecycle management. By combining an iterative agile approach to development with DevOps practices such as continuous integration and continuous delivery, organizations can achieve more streamlined and efficient development processes. This combination not only speeds up delivery cycles, but also ensures that systems are constantly updated and meet user needs and market conditions.

3. Flexible tasks in large enterprises

Adopting Agile at scale comes with unique challenges, especially when coordinating multiple teams and maintaining a unified vision across large projects. Solutions such as the Scaled Agile Framework (SAFe), and similar platforms, help large organizations effectively scale agile practices, ensuring that the benefits of agility and responsiveness maintained as the project size increases.

These discussions illustrate the current evolution of agile methodologies and their practical application in the development of modern information systems. Each of these topics—adapting to organizational needs, integrating with DevOps, and scaling across large enterprises—demonstrates the versatility and effectiveness of Agile in meeting the complex challenges of today's technology environment. This versatility is critical to developing, deploying, and maintaining reliable information systems that can adapt to the dynamic nature of technology and business.

Its iterative nature, combined with a strong focus on stakeholder engagement and flexibility, makes it exceptionally suitable for projects where requirements and goals may change based on new information and user feedback. These qualities are necessary to adapt to new technologies and market conditions, thereby ensuring the continued relevance and success of information systems.

Scrum

Scrum is a subset of agile methodologies designed specifically to manage the development of complex software and products. It uses fixed-duration iterations called sprints, with frequent reassessments and refinements to plans.

Scrum is a platform that helps teams work together to develop, deliver, and maintain complex products through iterative and phased methods. It is well-structured in its implementation, yet flexible enough to adapt to different projects. Scrum is especially known for its roles, rituals, and time frames that define the work process, such as daily meetings, sprint reviews, and sprint planning sessions.

Key principles and practices (Fig. 1.5).

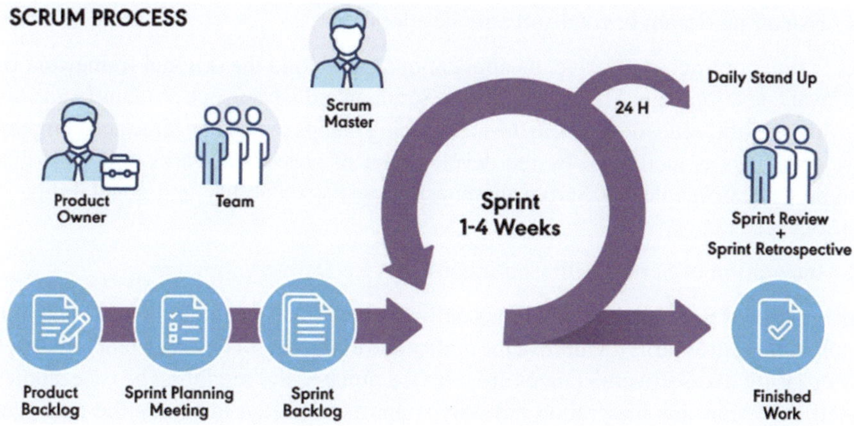

Fig. 1.5 Scrum development process

1. Scrum Roles
   - The framework defines three main roles: the product owner, who determines the details of the product and the order of work Scrum Master, who leads the team and monitors compliance with Scrum rules; and the development team, a self-organizing group responsible for improving the quality of the product.
2. Sprint
   - Sprints are a key element of Scrum, usually lasting two to four weeks, during which a useful product is created. Sprints are designed to create a rhythm of delivery, reflection, and adaptation.
3. Ceremonies
   - Scrum includes special ceremonies to ensure regular progress and continuous improvement. These include daily stand-ups, sprint planning meetings, sprint reviews, and retrospectives, each serving a specific purpose in the product development lifecycle.

Adaptation of Scrum in projects on traditional information systems

1. Scrum's flexibility allows it to be adapted to different types of projects, including traditional information systems. Scrum can be tailored to the specific requirements of non-software projects, improving project dynamics and deliverables. Their research shows that adapting Scrum to different organizational environments can significantly improve team collaboration and accelerate project delivery, demonstrating that Scrum's principles of iterative development and continuous feedback are beneficial for everyone.

2. Extending Scrum beyond software development

As Scrum continues to evolve, its applications go beyond the original framework of software development. The application of Scrum in industries such as manufacturing, healthcare, and education, where the methodology helps to manage large and complex projects. This is facilitated by the development of new frameworks and tools that support the principles of Scrum, demonstrating its scalability and adaptability in various areas.

3. Integration of Scrum with Continuous Service Delivery Practices

Integration of Scrum with continuous delivery methods is becoming more common. This combination aims to improve the agility and efficiency of development processes by ensuring that software updates are released quickly and reliably. The case studies cited show that this integration not only supports but often improves the pace and quality of development, which is important for organizations that need to adapt quickly to changes in technology or market demands.

Scrum's emphasis on agility, regular feedback, and iterative progress makes it ideally suited to manage the information systems development lifecycle in an ever-evolving technology environment. Its adaptable structure helps organizations respond quickly to changing market demands and technological advances, ensuring the reliability, relevance, and efficiency of the systems they develop.

Lean Development

The Lean methodology has its origins in manufacturing practice and is widely applied in a variety of industries, including software development, to improve processes and optimize productivity.

The methodology emphasizes waste reduction, efficiency, and the constant pursuit of creating value for customers. It is based on basic principles aimed at improving the quality of processes by eliminating non-value-added activities, i.e., waste.

Key principles and practices (Fig. 1.6).

1. Definition of value

The first step in Lean is to determine what value means to the customer—what they are willing to pay for. At this stage, all actions are coordinated with the needs of the client and the efficiency and purposefulness of the process are ensured.

2. Value Stream Mapping

It is crucial to plan all stages of the delivery process and identify the places where the waste originates. This process involves analyzing the flow of materials and information to identify inefficiencies.

3. Creating a flow

Once the waste is removed, the next steps will ensure a clear sequence of value-adding activities, which reduces cycle times and costs, improves quality, and ensures that value is delivered to the customer smoothly.

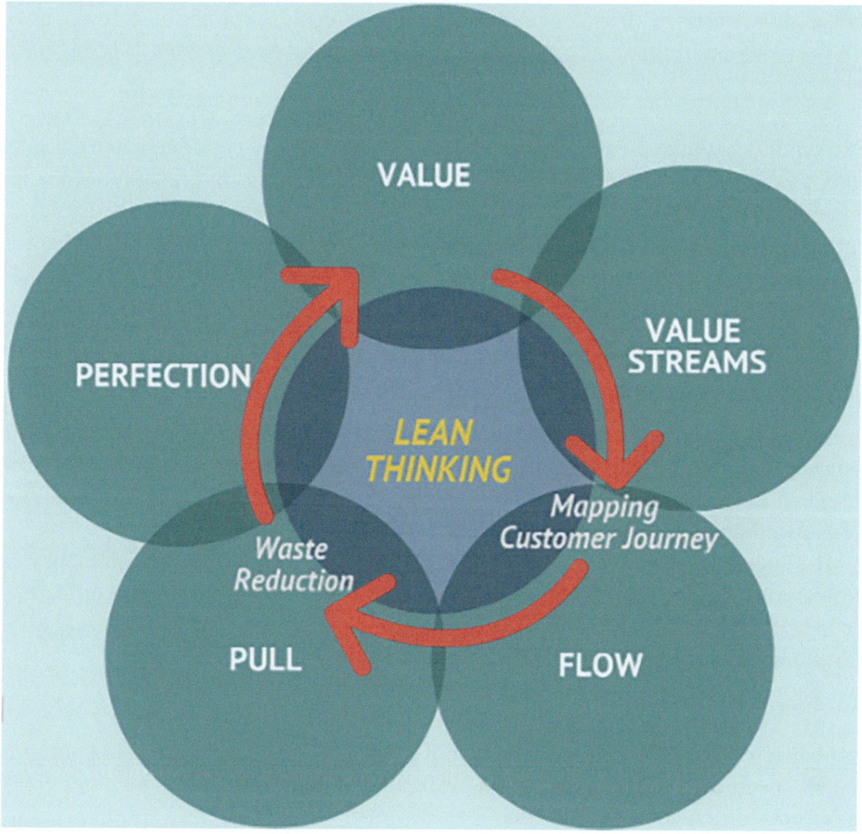

Fig. 1.6 Lean development process

4. Establishing Communication

Lean promotes systems based on information extraction, in which production is based on consumer demand rather than on the promotion of products to the market. This reduces the cost of inventory and increases the speed of response.

5. Striving for perfection

The goal of the methodology is continuous improvement. Striving for excellence means constantly looking for ways to improve and get closer to providing maximum value to customers without loss.

From recent research

1. Lean for improving IS efficiency and sustainability

The main goal of Lean is to improve the efficiency and sustainability of information systems. By applying Lean principles, organizations can optimize resource allocation and reduce waste in digital workflows, which is critical for the development of efficient and environmentally responsible systems.

2. Hybrid approaches that combine Lean manufacturing and flexibility

The synergy between Lean and agile methodologies creates a comprehensive framework that leverages the strengths of both methodologies. This hybrid approach helps organizations remain agile while increasing efficiency, especially for software development projects where dynamic change is common. Integrating Lean waste reduction strategies with Agile iterative development cycles contributes to a more agile and productive project management environment.

3. Overcoming Challenges in Lean IT Operations

There are practical challenges associated with implementing Lean in IT, such as overcoming resistance to change and the difficulty of applying lean principles to non-physical value streams. Strategies are proposed to address these challenges, such as enhanced training, better communication of the benefits of Lean, and modified tools specifically designed for the IT environment.

The methodology offers a systematic approach to improving the efficiency of project management. By focusing on value and continuously improving processes, Lean helps organizations adapt to technological changes and maintain a competitive advantage in their industry.

Kanban

Kanban is a visual project management tool developed by the Japanese manufacturing sector, specifically Toyota, and adapted for software development, and other areas, to improve efficiency and flexibility.

Kanban is a lean approach to work management pattern emphasizes getting work done on time without overburdening team members. In Kanban, work items are visually displayed on the board, allowing team members to see the status of each item of work at any time.

Key principles and practices.

1. Visualization of work

With a Kanban board, tasks are visualized to show progress at various stages of the development process, usually in columns such as "To Do," "In Progress," and "Done." Such visualization helps to identify bottlenecks and track the progress of work in real time.

2. Limitation of the number of unfinished works

A core principle of Kanban is to lzmit the work-in-progress (WIP), to prevent overload and focus on completing tasks before starting new ones. This helps to improve workflow and shorten cycle times.

3. Workflow Improvement

By managing workflow, Kanban helps teams work more smoothly and predictably. Improving workflow involves adjusting the scope, type, and speed of work, ensuring that tasks are completed from start to finish with as few interruptions as possible.

Key findings from recent research

1. Efficiency and adaptability in software development

Kanban increases agility in software development teams. The demonstrable nature of the methodology and the emphasis on limiting work-in-progress help teams quickly adapt to changes in project requirements and priorities, thereby maintain a high level of productivity and agility.

2. Scaling Kanban for larger projects

It is proposed to integrate Kanban with other flexible platforms, such as Scrum, to effectively manage larger, more complex projects. This approach preserves the advantages of Kanban in flexibility and flow control, while solving the problems of scaling and coordination in large teams.

3. Continuous improvement and training

Kanban's role in fostering a culture of continuous improvement and learning in teams. Using Kanban, teams can analyze their workflows, identify gaps, and implement improvements regularly. This continuous process of evaluation and adaptation is critical to ensure that teams remain competitive in a rapidly changing technology environment.

Kanban offers a comprehensive platform for information system lifecycle management, especially suitable for environments where flexibility and continuity of supply are valued. The principles of visual management, work-in-progress constraints, and workflow efficiency make it a powerful tool for increasing productivity and adaptability in the face of changing technology requirements.

DevOps

DevOps, which combines development (Dev) and operations (Ops), brings together software development, IT operations, quality assurance, and security to improve collaboration and productivity, allowing for faster product adoption.

DevOps is an approach that aims to bridge the gap between development, quality assurance, and IT operations, with a focus on communication, integration, automation, and collaboration. The goal is to streamline the development process and reduce the time required for deployment, while ensuring high quality and reliability.

Key principles and practices (Fig. 1.7).

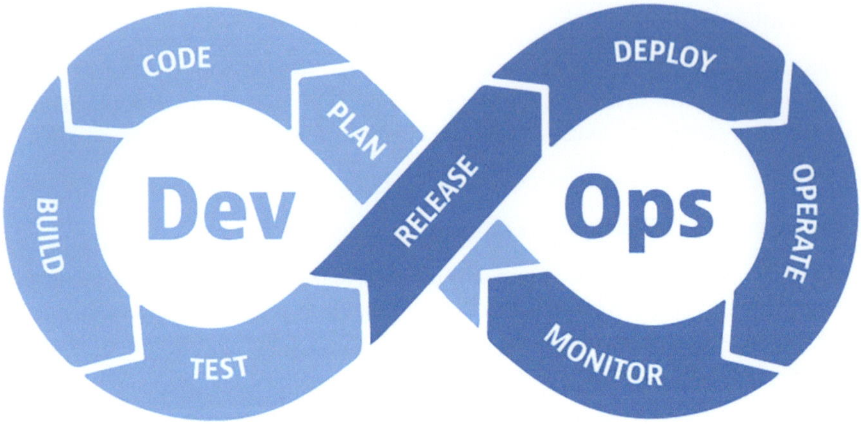

Fig. 1.7 DevOps development process

1. Continuous Integration/Continuous Deployment (CI/CD)

CI/CD is central to the DevOps approach, which involves regularly integrating code changes into a shared repository, followed by automated testing and deployment. This helps to identify and eliminate defects at an early stage, reduce the backlog, and improve the quality and speed of development.

2. Automation

Automation in DevOps encompasses testing, deployment, and configuration of the environment, reducing the need for manual intervention and improving speed and reliability. Automation helps eliminate repetitive tasks, allowing teams to focus on core development activities.

3. Monitoring and feedback

Regular monitoring of applications and infrastructure is vital in DevOps to ensure performance and reliability. Feedback mechanisms are integrated into different stages of the lifecycle to continuously improve processes and the final product.

Key findings from recent literature

1. Improve IT service productivity with DevOps

Significant improvement in operational efficiency and customer satisfaction as a result of the implementation of DevOps practices. Continuous delivery and automated testing, which are central to DevOps, help organizations shorten the development cycle and improve the reliability of IT services, which ultimately improves the customer experience.

2. DevSecOps: Integrating Security into DevOps

A DevSecOps concept that integrates security practices directly into DevOps workflows. This proactive approach ensures that security considerations are taken into account from the outset of development, resulting in more secure software solutions. By making security an integral part of the development and implementation process, organizations can more effectively mitigate risks and meet security standards without sacrificing performance.

3. Overcoming the challenges of implementing DevOps in enterprises

Challenges that large organizations face when adopting DevOps include cultural resistance, compatibility with legacy systems, and challenges related to scaling. Targeted strategies are proposed, such as developing specialized DevOps platforms, conducting intensive training, and using specialized tools to facilitate DevOps adoption at scale. These strategies are designed to help large enterprises overcome obstacles and take full advantage of DevOps to increase agility and efficiency.

DevOps is a modern paradigm that empowers organizations to develop, implement, and maintain software and services at high speed. This methodology supports information systems lifecycle management by promoting a culture of collaboration and shared responsibility, which is essential in a rapidly changing technology landscape.

Scaled Agile Framework (SAFe)

SAFe helps large organizations implement agile methodologies across teams and projects by facilitating alignment, collaboration, and implementation across multiple agile teams.

SAFe is a set of organization and workflow templates designed to guide businesses in scaling lean manufacturing practices and agile development. SAFe is designed for use in complex systems and large-scale software environments where multiple teams must coordinate in order to operate smoothly and efficiently.

Key principles and practices.

1. Centralized decision-making with decentralized execution

SAFe promotes centralized strategy and decentralized execution by enabling rapid decision-making at lower levels without compromising the strategic alignment provided by higher levels of governance.

2. Built-in Quality

Quality is considered at every stage of the development process, ensuring that each add-on meets quality standards and does not require extensive rework.

3. Program Growth Planning

Planning is an important practice in SAFe, involving all team members in planning work for the next 8–12 weeks. This in-person event promotes better alignment and coordination between teams.

Key findings from recent research

1. Increasing the flexibility of the enterprise

SAFe significantly increases enterprise agility by integrating customer feedback and continuous improvement processes. This helps large organizations reduce the time required to get their product to market and adapt faster to market changes and customer needs.

2. Leadership and Culture Changes

Successful implementation of SAFe depends to a large extent on the commitment of management and the transition to a lean and agile culture in the organization. The transition involves the introduction of flexible values at all levels of the company, which is necessary to overcome resistance and ensure the effectiveness and sustainability of agile transformation.

3. DevOps Integration

Integrating DevOps practices into a secure environment can further enhance organizational ability to deliver services. By implementing automation and continuous delivery, security professionals can improve the transfer of value across conveyors and achieve higher efficiency and reliability in product delivery.

SAFe offers a structured approach to scaling agile development, which is essential for managing large projects involving multiple teams. It aligns strategy and execution, enabling large enterprises to innovate faster, deliver services more efficiently, and take advantage of agile development across complex systems.

Waterfall

The Waterfall methodology is a traditional, linear and sequential approach to software development, in which each stage must be completed before the next one begins. Waterfall emphasizes the importance of careful planning, documentation and structured walkthroughs of the stages, ensuring that each step is fully implemented before moving on.

Waterfall is one of the earliest and most widely used software development methodologies. The process is divided into separate stages, each of which has specific results and goals. These phases typically include requirements elicitation,, design, implementation, testing, deployment, and maintenance. The Waterfall model is simple and provides a solid structure for projects with stable requirements and little need for change throughout the development process.

Key principles and practices.

1. Successive stages

The Waterfall methodology strictly adheres to a linear sequence. Each stage must be completed in its entirety before the next can be started, ensuring that teams are methodical and that there is no development work until the design is complete. This minimizes the risk of design errors during the development phase.

2. Comprehensive documentation

Documentation is crucial in Waterfall projects. Requirements, design plans, and test cases are carefully documented before implementation begins, ensuring a clear understanding of the project at every stage. This makes it easier to manage and transfer knowledge between teams and stakeholders.

3. Defined milestones and deliverables

In the "Waterfall" model, special attention is paid to stages. There are well-defined points at which deliverables are created, such as a fully documented set of requirements or a completed project plan. These milestones serve as markers to measure progress and align team goals.

Key findings from recent research

1. Suitability for fixed-scale projects

Waterfall is effective in environments where the requirements are well understood and unlikely to change. This method provides a structured and predictable process that is easier to manage in industries such as public administration and healthcare, where compliance and documentation quality are paramount.

2. Problems of adaptation to changes

One of the main drawbacks of the Waterfall model is its inflexibility. Once the stage is complete, reviewing is often costly and time-consuming. In projects with changing requirements, this rigidity can lead to delays and increased costs, making them less suitable for dynamic or iterative development environments.

3. Improved risk management through structured processes

Waterfall provides a clear and structured approach to risk management. Risks are identified at an early stage of the project, and mitigation strategies are developed at the planning stage. This proactive risk management allows teams to prevent major disruptions in the development process.

Waterfall remains a valuable methodology in certain project environments, especially where requirements are stable and thorough documentation is required. However, its rigidity and linear nature make it less suitable for projects that require frequent iterations or changing goals. Despite these limitations, Waterfall continues to support effective information systems lifecycle management, especially in industries with low resilience to change and high requirements for accuracy and compliance.

1.7 Comparative Analysis

The evaluation of methodologies will be based on efficiency, flexibility and compliance with business strategies.

Efficiency measures how well a methodology delivers successful outcomes, such as meeting deadlines and meeting user requirements.

Flexibility refers to the ability to adapt to changing requirements throughout the project lifecycle.

Business strategy alignment assesses how well a methodology aligns with an organization's broader goals, such as innovation or operational efficiency.

These parameters help organizations choose a methodology that delivers results, adapts to change, and supports strategic goals.

1. Agile

Efficiency: Agile is very effective in dynamic environments where constant feedback and rapid iterations are required. It promotes iterative development and adaptive planning, enabling continuous improvement.

Flexibility: Extremely flexible; Enables teams to respond quickly to change, adapt to new requirements, and prioritize tasks according to current business needs.

Alignment with business strategies: Aligns well with modern business strategies focused on innovation, customer feedback, and fast product delivery. Agile promotes collaboration and accelerates a product's time to market.

2. Scrum

Efficiency: Scrum is very effective in managing complex projects by breaking them down into small, manageable sprints. Its time-limited approach ensures rapid implementation and continuous improvement.

Flexibility: Provides flexibility during short sprints, but once a sprint starts, changes can only be made to the following one. This provides a balance of stability and adaptability.

Align with business strategies: Scrum's focus on generating small value increments makes it suitable for companies looking for quick wins and continuous operations. This ensures compliance through regular analysis and feedback.

3. Lean development

Efficiency: Lean effectively eliminates waste and streamlines processes, making it especially useful in manufacturing environments and software development. It is aimed at obtaining the maximum return with the minimum expenditure of resources.

Flexibility: Lean manufacturing is moderately flexible and allows to make changes to processes based on real-time feedback. However, it focuses on efficiency, which can slow down innovation in fast-growing industries.

Alignment with business strategies: Pairs well with businesses focused on improving operational efficiency, reducing costs, and maximizing customer value. Lean manufacturing is especially valuable in industries where minimizing waste and improving processes are critical.

4. Kanban

Efficiency: Kanban is effective in environments with a constant flow of work, such as service desks or operations teams. It visualizes workflows and helps manage production facilities and optimize the production process.

Flexibility Kanban is highly flexible: A high degree of flexible because it does not prescribe specific roles or iterations. This ensures a continuous flow and prioritization adjustment without disrupting the entire workflow.

Alignment with business strategies: Kanban is well-suited for businesses focused on continuous delivery and process optimization. This helps to reduce bottlenecks, better manage resources, and increase efficiency.

5. DevOps

Efficiency: DevOps is extremely effective in environments where continuous integration, delivery, and deployment are critical. It bridges the gap between development and operations, enabling faster releases and more reliable systems.

Flexibility: DevOps provides high agility by automating processes and ensuring continuous delivery of software. This promotes iterative improvement and rapid adaptation to change.

Alignment with business strategies: DevOps aligns with business strategies that emphasize speed, reliability, and continuous innovation. This supports the culture of collaboration and shared responsibility required for digital transformation initiatives.

6. SAFe

Efficiency: SAFe is very effective for large enterprises that require coordination between multiple teams working on different parts of the same project. It scales agile principles to ensure consistency across the organization.

Flexibility: SAFe retains some flexibility but is more structured than Agile, striking a balance between rapid iterations and the need for enterprise-level planning and coordination.

Alignment with business strategies: SAFe pairs well with large organizations looking for flexibility at scale. It supports business strategies related to large and complex projects that require collaboration between different departments, but at the same time focuses on generating quick profits.

7. Waterfall

Efficiency: Effective in projects with well-defined and stable requirements, such as regulatory or compliance-oriented initiatives. This structured approach ensures thorough documentation and planning.

Flexibility: Minimal flexibility due to its linear and sequential nature. Post-phase changes can be costly and time-consuming, making them unsuitable for projects with changing requirements.

Alignment with business strategies: Aligns with traditional business strategies that require clear project timelines, thorough documentation, and well-defined milestones. Best suited for projects where predictability and risk management are key.

Each methodology has its own strengths and weaknesses, with Agile, Scrum, and DevOps being the most agile for dynamic environments, while Waterfall and Lean provide more stability and structure for projects with well-defined requirements. SAFe bridges the gap for large enterprises that need flexibility at scale.

1.8 Other Tools and Technologies for Lifecycle Management

In addition to lifecycle methodologies and models, there are many tools and technologies that help manage the lifecycle of information systems. These tools focus on various aspects such as planning, development, deployment, monitoring, and maintenance. The key categories of tools and technologies used in information systems lifecycle management are:

1. Project Management Tools
   - Jira: Used for agile project management, bug tracking, and issue tracking. It helps teams plan, track, and release software;
   - Trello: A visual tool for organizing tasks and tracking progress using boards, lists, and cards;
   - Asana: Focused on project and task management, provides scheduling, deadline tracking, and team collaboration features.
2. Version control systems
   - Git: A distributed version control system that allows developers to track changes to source code during the development process;
   - GitHub/GitLab/Bitbucket: Platforms that provide cloud hosting for Git repositories, as well as features for collaboration, code review, and project management.
3. Continuous Integration/Continuous Deployment (CI/CD) Tools
   - Jenkins: An open-source automation server used to automate the CI/CD pipeline, enabling continuous integration, testing, and deployment of code;
   - CircleCI: A CI/CD tool that automates the process of building, testing, and deploying software projects;
   - Travis CI: A cloud-based CI service used to build and test software projects hosted on GitHub.
4. Configuration Management and Automation Tools
   - Ansible: An automation tool that provides easy configuration management, application deployment, and task automation across distributed systems;
   - Puppet: Automates the provisioning, configuration, and management of servers and applications, ensuring consistency across environments;

- Chef: Automates infrastructure configuration, ensuring that servers and applications are configured according to defined policies.

5. Cloud Platforms and Infrastructure Management Tools

   - Amazon Web Services (AWS): Provides cloud services for computing, storage, networking, and database management. CloudFormation help automate provisioning;
   - Microsoft Azure: Offers cloud services with a wide range of tools for managing information systems, from databases to virtual machines and serverless functions;
   - Terraform: An open-source infrastructure as a code (IaC) allows to define and provide data center infrastructure using a declarative configuration language.

6. Monitoring and logging tools

   - Prometheus: A monitoring tool that collects metrics from applications and services, allowing to monitor system performance and availability in real time;
   - Grafana: Works with Prometheus to visualize monitoring data and create dashboards to track key performance indicators;
   - Splunk: A powerful log analytics tool that collects and indexes log data, helping organizations troubleshoot and monitor systems in real-time.

7. Test automation tools

   - Selenium: A widely used open-source tool for automating web browser interactions, commonly used for functional and regression testing;
   - JUnit: A testing platform, used primarily in Java projects for unit testing, to ensure that individual components of the system work as expected;
   - TestNG: Another testing platform that provides more powerful and flexible testing capabilities than JUnit and supports different types of testing such as unit, integration, and end-to-end testing.

8. Collaboration and documentation tools

   - Confluence: A collaboration tool used to document, share knowledge, and track projects. It integrates with Jira to improve project management;
   - Slack: A communication platform that allows teams to collaborate, share files, and integrate with other tools like GitHub, Jira, and Trello to ensure seamless workflows;
   - Microsoft Teams: A collaboration, video conferencing, and file-sharing tool often used in conjunction with Microsoft Office tools.

9. Enterprise Resource Planning (ERP) tools.

   - SAP ERP: Enterprise software that manages business operations and customer relationships, often integrating with other systems to support lifecycle management;

   - Oracle ERP Cloud: A suite of finance, HR, and supply chain applications that help streamline enterprise operations.

10. Database Management Tools
   - MySQL Workbench: A visual tool for managing and modeling MySQL databases. It provides tools for designing, developing, and administering databases;
   - pgAdmin: A comprehensive PostgreSQL database management tool that offers features such as query execution, database modeling, and monitoring;
   - MongoDB Atlas: A cloud-based database management system for managing MongoDB databases that allows to scale, monitor, and backup.
11. Enterprise Architecture Tools
   - TOGAF: A platform that provides methods and tools for designing an enterprise architecture and aligning it with business strategy;
   - ArchiMate: An enterprise architecture modeling language that offers a unified way to describe and visualize enterprise systems and their relationships;
   - Sparx Systems Enterprise Architect: A simulation tool used to design enterprise architectures, systems, and software, providing support for UML and other modeling standards.
12. Business Process Management (BPM) Tools
   - Camunda: A document and decision-making automation platform that helps you design, execute, and optimize business processes.
   - Bizagi: A BPM platform that offers tools for modeling, automating, and optimizing business processes, supporting integration with other information systems.

These tools and technologies help manage various aspects of the information systems lifecycle, from planning and development to deployment, monitoring, and continuous improvement. They provide critical support to keep systems running efficiently, remain agile, and align with changing business strategies.

1.9 Technology Landscape

The field of information systems planning is undergoing rapid changes due to the emergence of new technologies and changes in business priorities. While many trends, such as AI integration, automation, and cloud computing are already in place, their impact is expected to increase significantly. Let us look at the emerging trends and forecasts, and how they will change the methodologies, tools and strategies used in information systems planning. As organizations face increasing complexity and new demands, understanding future developments will be critical to remaining competitive and responding to technological innovations.

1. Strengthening the integration of artificial intelligence and machine learning

Future developments in information systems planning are likely to lead to a significant increase in the use of AI and machine learning to automate decision-making and optimize resource allocation. AI-driven systems can predict project bottlenecks, streamline workflows, and provide real-time insights based on historical data. These technologies will enable more dynamic and adaptive planning, in which systems will continuously learn from new data, adjusting plans according to changes in technology, markets, or business needs.

Impact on methodologies

Methodologies such as Agile and DevOps can evolve to include AI-driven feedback loops that enable teams to make data-driven decisions faster. Tools can automate sprint planning, task prioritization, and even autonomously generate code or perform testing. This will expand the capabilities of project management tools from simple tracking to active participation in planning and execution processes.

2. Transition to autonomous systems and automation

As more systems become autonomous, the need for human intervention in routine planning and execution tasks will decrease. Automation will dominate both the operational and development phases, thanks to self-correcting systems that adjust processes in real time based on pre-set performance thresholds. For example, cloud infrastructure can manage resources on its own, scaling up or down, based on demand, without manual control.

Impact on methodologies

Traditional rigid methodologies like Waterfall will become even less relevant as real-time adjustments require more flexible approaches like Agile, Scrum, and Kanban. DevOps is likely to evolve into a "NUPS" for certain industries where operations are fully automated, allowing development teams to focus on innovation.

3. Blockchain for planning and management.

Blockchain technology can become a central component of information systems planning, providing secure, transparent, and immutable records of changes to the system, resource allocation, and workflows. This will improve project management and accountability, especially in distributed and cross-border projects where trust and compliance are critical.

Impact on methodologies

Blockchain integration can formalize and simplify project auditing and management. Agile methodologies can use blockchain to track user stories, tasks, and sprint cycles, while large-scale platforms like SAFe can use it to coordinate collaboration across multiple teams with transparent tracking of system contributions and updates.

4. Paying more attention to cybersecurity and data privacy

With data breaches on the rise and data protection regulations (such as GDPR) tightening, security will be prioritized early in the planning phase. Future methodologies will include security protocols at every stage, similar to how DevSecOps currently integrates security into the DevOps cycle.

Impact on methodologies

DevOps will scale toward DevSecOps, which will ensure that security considerations are taken into account in all aspects of planning and implementation. Agile teams may need to add dedicated security sprints to maintain compliance with the new rules. This can complicate existing workflows, but it can also make systems more resilient and secure from the ground up.

5. Development of quantum computing

As quantum computing advances, information systems planning will need to consider quantum-enhanced technologies that can exponentially accelerate data processing, encryption, and task optimization. Quantum resource integration planning could become the new standard in high-performance computing environments, enabling more complex models and simulations to be created early in development.

Impact on methodologies

Methodologies such as Waterfall, which are traditionally used for large projects that require careful pre-planning, can be revived in a quantum computing environment, where simulation and pre-execution planning can reduce risk. Agile methodologies can also use quantum computing to perform a complex data analysis on each iteration, which speeds up decision-making.

6. Remote and decentralized workforce

The shift to remote work, accelerated by the COVID-19 pandemic, will continue to impact the planning and management of information systems. Tools that facilitate seamless remote collaboration, such as cloud-based IDEs, communication platforms, and real-time communication systems, will be indispensable when planning for the future.

Impact on methodologies

The Agile and Scrum platforms will continue to evolve to enable fully decentralized teams. AI-integrated collaboration tools can automate time zone planning, resource allocation, and sprint management across distributed teams. Methodologies such as SAFe can evolve to include specific guidelines for managing decentralized development and operations across global teams.

7. Economic pressures and budget constraints

The economic downturn and increased competition in the global market will lead to tighter budget constraints and resource management at the planning stages. This is likely to force enterprises to adopt cost-optimized solutions, including open-source technologies and pay-as-you-go cloud services.

Impact on methodologies

Under these conditions, the Agile and Lean methodologies will be successfully applied, given their focus on maximizing value with limited resources. Enterprises will increasingly adopt Lean to reduce waste and inefficiencies, while the iterative nature of Agile will allow for more precise budget control and adjustments as projects progress. Moving to a cost-effective cloud infrastructure will also push companies to focus on technologies that provide scalability without excessive upfront investment.

Future developments in information systems planning are heavily influenced by new technologies such as artificial intelligence, quantum computing, and blockchain, as well as economic and environmental pressures. Agility, automation, and security are central to the evolution of planning methodologies, with Agile, DevOps, and hybrid models becoming increasingly dominant due to their ability to adapt to constant change. Economic factors are driving the further adoption of cost-effective and sustainable practices in both technology selection and process management.

1.10 Conclusion

This Chapter reviewed the existing models, methods and techniques to build complex and large-scale digital products; it identified a set of useful patterns and helpful practices for their sustainable development in crisis conditions.

References

1. Zykov, S.V.: Crisis management for software development and knowledge transfer issue 61: Springer series in smart innovation, systems and technologies. Springer, Switzerland (2016)
2. Rocha, Á., Adeli, H., Dzemyda, G., Moreira, F., Poniszewska-Marańda, A. (Eds.): Good practices and new perspectives in information systems and technologies: WorldCIST 2024, Volume 1. Springer, Cham. https://doi.org/10.1007/978-3-031-60215-3. Softcover ISBN 978-3-031-60214-6; eBook ISBN 978-3-031-60215-3
3. Buyya, R., Broberg, J., Goscinski, A.M.: Cloud computing: Principles and paradigms. Wiley, Hoboken (2018). ISBN 978-0470940099
4. Paspallis, N., Raspopoulos, M., Barry, C., Lang, M., Linger, H., Schneider, C., (eds.): Advances in information systems development: Methods, tools and management. In: Conference Proceedings, (2018)

5. Kumar, S., Verma, A.K., Mirza, A.: Digital transformation, artificial intelligence and society: opportunities and challenges. Springer, Singapore (2024). https://doi.org/10.1007/978-981-97-5656-8. Hardcover ISBN 978-981-97-5655-1; Softcover ISBN 978-981-97-5658-2; eBook ISBN 978-981-97-5656-8
6. Dey, N., Das, H., Naik, B. and Behera, H.S. (eds.): Big data analytics for intelligent healthcare management. Academic Press, Amsterdam (2019). ISBN 978–0128181461

Chapter 2
Architecting Transaction Management: From Monolith to Microservices

Abstract This chapter illustrates the techniques to improve transaction management in large and, therefore, complex IT infrastructures. The complexity as a potential crisis trigger, may result from bulky organizational structure, spatial and business diversity, or another reason. The approach is reinforced by formal optimization modeling and verified by a user survey. DevOps techniques and practices assist in promoting the approach. Following the IT Crisisology framework, this methodology incorporates a guideline on the complete triad of formal models, agile methods, and supporting tools.

2.1 The Monoliths

2.1.1 What is a Monolith?

A monolith application is a single archived, self-contained, and independent application that runs on a single platform. It consists of a database, client-side user interfaces, and a server-side application in one single software executable program. While it is just a single and independent unit of work, most of the time there are multiple teams of developers working on the same application. Each of the team is responsible for the individual components they are working on (see Fig. 2.1).

The figure illustrates the basic architecture of a monolithic application. The problem here is that as the size and complexity of the application is increased, the coordination and communication of the individual team is not scaled. Every time a new feature is added, or a change is done in the currently deployed code, the entire application has to be rebuilt, redeployed, and tested.

Although the size and complexity of the monolithic application grew, the communication and coordination costs of the individual teams working on the application did not scale. Every time an individual team needs to make a change, the entire application has to be rebuilt, retested, and redeployed. The concept of a microservice originally crept into the software development community's consciousness around

S. V. Zykov, *IT Crisisology Patterns and Practices*, Smart Innovation, Systems and Technologies 449, https://doi.org/10.1007/978-981-95-1260-7_2

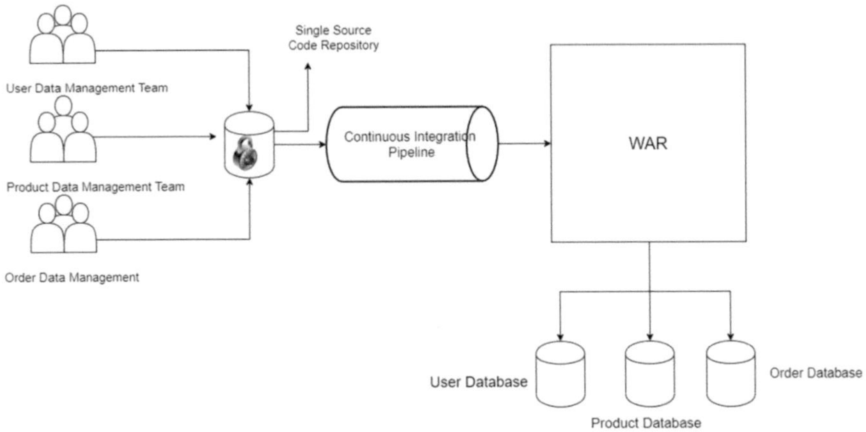

Fig. 2.1 Basic architecture of a monolithic application

2014 and was a direct response to many of the challenges of trying to scale both technically and organizationally large, monolithic applications.

When the application was relatively small, its monolithic architecture had the following benefits:

- Easy to make radical changes to the application—The developers could change the code and the database schema, build, and deploy
- Straightforward to test—The developers wrote end-to-end tests that launched the application, invoked the REST API, and tested the UI with Selenium
- Straightforward to deploy—All a developer had to do was copy the WAR file to a server that had Tomcat installed
- Easy to scale—can run multiple instances of the application behind a load balancer.

2.1.2 Living with Monoliths

As a company becomes more successful, the size of the development team steadily grows. Not only does this increase the growth rate of the code base, it also increases the management overhead.

The problem with monoliths is that they become too complex as they mature and age. It is too large for any developer to fully understand the resulting bugs and correctly implementing new features becomes difficult and time consuming. Deadlines are missed.

If the code base is difficult to understand, developers cannot make the correct changes. Each change makes the code base incrementally more complex and harder to understand.

Developers find their day-to-day tasks slow as the system contains overwhelming complexity. The larger the application the more time it takes to load into the IDEs of the developer and requires more time to debug as well as to see feature changes. As a result, the edit-build-run-test loop takes a long time, which hampers developer productivity.

The operations team also has scaling problems because there can be a module resources conflict. As the load on your grows application, the first performance issue is the application slows down due to the volume of requests coming in.

While working in a monolith there is no liberty to choose the technology stack as per the component requirement, and the developers must code the component in the same language used to code other components. The monolithic architecture makes it difficult to adopt new frameworks and languages. It would be extremely expensive and risky to rewrite the entire monolithic application so that it would use a new and presumably better technology. Consequently, developers are stuck with the technology choices they made at the start of the project. Quite often, they must maintain an application written using an increasingly obsolete technology stack.

2.2 The Microservice Architecture

With a monolith, it is easy to start development. But as the system becomes bigger and more complicated over time, it becomes very difficult to be agile and productive. Various qualities, such as scalability and agility, need to be maintained for the lifetime of the application. This becomes complicated as the system needs to be continuously updated as requirements keep changing and evolving over time.

Any business-domain service may be implemented as a single process, accessed via an API over the network, completely self-contained and opaque, independently developed and deployed without regard to any other services even in the same application, and usually built and deployed using automated processes. An implementation is not a microservice even if it is deployed in a Docker container unless it also meets this definition. A monolithic application in a Docker container is not a microservice.

2.2.1 Decomposition of Application in Microservices

There are various ways to decompose an application [2]. This section discusses two different ways of breaking down an application (see Fig. 2.2).

Scale cube and microservices

1. **X-axis Scaling**: This is achieved by cloning the application and data along multiple servers. A pool of requests is sent to a load balancer and the requests are delegated to the servers. Each of the servers has the latest version of the application and has access to all the data required. In this regard, it makes no difference

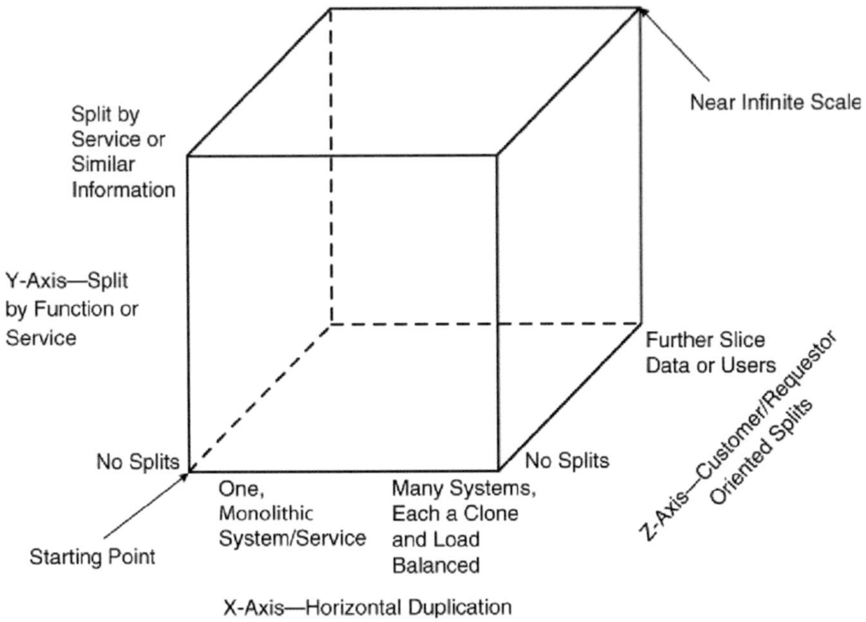

Fig. 2.2 Scale cube and microservices

which server fulfills the request. Rather, this is about how many requests are fulfilled at any this time. It is easy to scale along the X-axis as the number of requests increases. The solution is simply adding additional clones. However, this type does not scale with an data increase. Moreover, it also does not scale when there is a large variation in the frequency of any type of request because all the requests are handled in an unbiased way and are allocated to servers in the same way.

2. **Z-axis Scaling**: This scaling is obtained by splitting the request based on certain criteria. It is different from X-axis scaling as that the servers are responsible for different types of requests. Usually, the servers have the same copy of the application, but some can have additional functionalities depending upon the requests expected. Z-axis scaling assists with fault isolation and transaction scalability. This type provides extra function to certain customers groups. Additionally, new functionality is tested for a small group of customers to minimize risk.
3. **Y-axis Scaling**: This dimension means splitting application responsibility. The separation can be done either by (a) data, (b) operations on the data, (c) combination of (a) and (c). The respective methods can be referred to as resource-oriented or service-oriented splits. While the X-axis or Z-axis splits duplicate of the same functions, Y-axis duplicates certain function along servers. The major advantage is that each request is scaled and handled differently according to its importance. As the logic and the associated data are separated, developers can focus and work on a small section at a time. This increases both productivity and agility.

Additionally, any component fault is isolated and handled gracefully without affecting the rest of the application. However, scaling along the Y-axis can be costly compared to the other dimensions.

Modular application development is key to building successful large-scale complex applications to be decomposed, developed, and understandable by multiple developers.

The microservice architecture uses services as a unit of modularity. A service having a bounded context, which means a dense boundary that cannot be violated. The service is totally encapsulated and abstracted by a service API. As a result, it preserves service modularity.

Each microservice data is private and accessible via its API only.

There are different ways to keep service data private [2, 3]. For a relational database, the options are:

- Private-tables-per-service—each service owns a set of tables only accessible by that service
- Schema-per-service—each service has a database schema private to that service
- Database-server-per-service—each service has its own database server.

Private tables-per-service and schema-per-service have the lowest overhead. Using a schema per service is appealing since it makes ownership clearer. High-throughput services might need their own database server.

At the high level, both SOA and the microservice architecture are architectural styles that structure a system as a set of services.

SOA and the microservice architecture usually use different technology stacks. SOA applications typically use heavyweight technologies such as SOAP and other web service standards. They often use an ESB, a smart pipe that contains business and message processing logic to integrate the services. Applications built using the microservice architecture tend to use lightweight, open-source technologies. The services communicate via dumb pipes, such as message brokers or lightweight protocols like REST or gRPC. See Table B.1 of Appendix B.

Another key difference between SOA and the microservice architecture is the size of the services. SOA is typically used to integrate large, complex and monolithic applications. Although services in a microservice architecture are not always tiny, they are almost always much smaller. As a result, an SOA application usually consists of a few large services, whereas a microservice-based application typically consists of dozens or hundreds of smaller services.

2.3 Quality Factors of Microservices

2.3.1 Introduction

The most uncompromising factor of software development is quality. Regarding SOA, complexity increases with more requirements and time. So, for services developed as applications, quality becomes a critical factor. To overcome this, quality models are introduced over time.

The internal quality factors expose the design time factor which needs to be considered during the application design phase. Some internal quality factors are cohesion, coupling, granularity, and autonomy. Also, the internal quality factors are the characteristics of the software artifact produced at the end of SDLC, and these characteristics can only be measured at the end of the cycle.

Studies have been conducted to evaluate factors which affect microservice quality [5].

Research focused on the service granularity coupling (see Table B.2 of Appendix B).

2.3.2 Metrics for Microservice Quality

This section describes modular architecture and object-oriented quality metrics refined for used in service-oriented systems [3–5].

Before introducing the metrics, let us define:

- The domain of a business with varied processes of $P = \{p_{1,} p_{2,}, \ldots\ldots, p_{3,}\}$.
- Service set of perceived application: $S = \{s_{1,}\ s_{2,} \ldots\ldots, s_{3,}\}$.
- For any service $s \in S$, the operation set by s is $O(s) = \{o_{1,} o_{2,} \ldots\ldots, o_3\}$ *and* $|O(s)| = O$.
- Set of Input and output messages $M(o)$ for an operation: $o \in O(s)$. So, set of messages for all operations: $M(s) = \cup_{o \in O(s)} M(o)$.
- Consuming service s $\in S$ is perceived by $S_{consumer}(s) = \{s_{c1,} s_{2,} \cdot, s_{nc}\}$; n_c is the number of consumers for the service.
- The service dependency or the producer of the service s $\in S$ depends on $S_{producer}(s) = \{s_{c1,} s_{2,} \ldots\ldots, s_{np}\}$; n_p is the number of producers.

Metrics to determine coupling:

$$SOCI(s) = number\_of\_operations\_invoked = |\{o_i \in s_i : \exists_{o \epsilon s} calls(o, o_i) \wedge s \neq s_i\}|$$

$$ISCI(s) = number\_of\_services\_invoked = \left|\left\{s_i : \exists_{o \epsilon s}, \exists_{o_i \epsilon s_i} calls(o, o_i) \wedge s \neq s_i\right\}\right|$$

$$MCI(s) = size\_of\_message\_required\_from\_other\_services = \\ = |\cup M(o_i) : (o_i \in s_i) \wedge s \neq s_i|$$

where,

- *SOCI*—Service Operation Coupling Index i.e., number of operations of the service invoked by the service
- *ISCI*—Inter Service Coupling Index i.e., number of services invoked by the service
- *SMCI*—Service Message Coupling Index i.e., total number of messages from the information model required by the operations
- $calls(o, o_i)$ represents call made from service 'o' to service 'o_i'.

Coupling is evaluated as the information entropy of a complex service component where entropy is calculated using various data such as total number of atomic components connected, total number of links to atomic components, and total number of atomic components:

$$Coupling(S_i) = p\sum_{j=1}^{n_s}(-loglog(P_L(j))) \\ = \frac{1}{n}\sum_{j=1}^{n_s}(-loglog(P_L(j)))$$

where

- n_s—total number of services connected
- 'n'—total number of services in the entire application
- $\mathrm{P} = \frac{1}{n}$—the probability of a service participating in any connection

Coupling is measured by using the dependency of a service operation with the other services. Additionally, the dependency is considered to have different weight value for each kind of operation such as Create, Read, Update, Delete (CRUD), and is based on the type of business entity involved. The weight is referred from the CRUD matrix constructed in the process:

$$coupling = \frac{dependency\_on\_business\_entities\_of\_other\_services}{number\_of\_operations\_and\_dependencies} \\ = \frac{\sum_{i \in D_s}\sum_{k \in O(s)} CCO(i, k)}{|O(s)|.K}$$

where

- $D_s = \{D_1, D_2, \ldots\ldots, D_k\}$—set of dependencies that a service has on other services
- $K = |D_s|$

- *CCO*—imaginary coupling between service operations obtained from business entities and business operations.

$$coupling = \frac{number\_of\_service\_connected}{number\_of\_services} = \frac{n_c + n_p}{n_s}$$

- n_c—number of consumer services
- n_p—number of dependent services
- n_s—total number of services

$$coupling = \frac{number\_of\_invocation}{number\_of\_servcies} = \frac{\sum_{i=1}^{|O(s)|}(S_{i,sync} + S_{i,async})}{|S|}$$

where

- $S_{i,sync}$—synchronous invocation of operation O_i
- $S_{i,async}$—asynchronous invocation in the operation O_i.

$$SFCI = \frac{number\_of\_operations\_using\_same\_message}{number\_of\_operations} = \frac{max(\mu(m))}{|O(s)|}$$

where

- *SFCI*—Service Functional Cohesion Index

If a service uses one message for all operations, it is considered as highly cohesive.

$$cohesion = \frac{n_s}{|M(s)|}$$

$$SIDC = \frac{number\_of\_operations\_sharing\_same\_parameter}{total\_no\_of\_parameters\_in\_a\_service} = \frac{n_{op}}{n_{pt}}$$

$$SIUC = \frac{sum\_of\_number\_of\_operations\_used\_by\_each\_consumer}{product\_of\_total\_number\_of\_consumers\_and\_operations}$$

$$SSUC = \frac{sum\_of\_number\_of\_sequentials\_operations\_accessed\_by\_each\_consumer}{product\_of\_total\_number\_of\_consumers\_and\_operations}$$

$$= \frac{n_{s0}}{n_c * |O(s)|}$$

where

- *SIDC*—Service Interface Data Cohesion
- *SIUC*—Service Interface Usage Cohesion
- *SSUC*—Service Sequential Usage Cohesion
- n_{op}—number of operations in the service which share the same parameter type
- n_{pt}—total number of distinct parameter types in the service
- n_{oc}—total number of operations in the service used by the consumers
- n_{s0}—total number of sequentially accessed operations by the client of the service.

The service capability granularity is calculated by the number of operations in a service and the data granularity by the number of messages consumed by the operations of the service [5].

$$SCG = number\_of\_operations = |O(s)|$$

$$SDG = size\_of\_messages = |M(s)|$$

where

- SCG—service capability granularity
- SDG—service data granularity

Also, it is evaluated as the ratio of squared number of operations of the service to the squared number of messages consumed by the service:

$$granularity = \frac{number\_of\_operations}{size\_of\_messages} = \frac{|O(s)|^2}{|M(s)|^2}$$

Reference perceived:

$$ODG = fraction\_of\_total\_weight\_of\_input\_and\_output\_parameters$$

$$= \left(\frac{\sum_{i=1}^{n_i o} W_{pi}}{\sum_{i=1}^{n_i s} W_{pi}} + \frac{\sum_{j=1}^{n_j o} W_{pj}}{\sum_{j=1}^{n_j s} W_{pj}} \right)$$

$$OFG = complexity\_weightage\_of\_operation = \frac{W_i(o)}{\sum_{i=1}^{|O(S)|} W_i(o)}$$

$$granularity = sum\_of\_product\_of\_data\_and\_functionality\_granularity$$

$$= \sum_{i=1}^{|O(S)|} ODG(i) * OFG(i)$$

where

- ODG—operation data granularity
- OFG—operation functionality granularity
- W_{pi}—weight of input parameter
- W_{pj}—weight of output parameter
- $n_i o$—number of input parameters in an operation
- $n_j o$—number of output parameters in an operation
- $n_i s$—total number of input parameters in the service
- $n_j s$—total number of the output parameters in the service
- $W_i(o)$—weight of an operation of the service
- $|O(S)|$—number of operations in the service.

ODG is evaluated in terms of number of input and output parameters and their type such as simple, user-defined, and complex.

OFG is defined as the level of logic provided by the operations of the services.

SOG is defined by the sum of product of data granularity and functionality granularity for all operations in the service.

Complexity depends on the degree of coupling for a service and evaluated as the fraction of its coupling to the total number of services, also measured as the fraction of total dependency weight of consumers upon the service to the total number of services [15].

$$RIS = \frac{IS(s_i)}{|S|}$$

$$RCS = \frac{coupling\_of\_service}{number\_of\_service} = \frac{CS(s_i)}{|S|}$$

where

- CS (s_i)—coupling value of the service
- $|S|$—number of services to realize the application
- $IS(s_i)$—importance weight of a service in the application

Complexity also depends on the granularity of service operations [10].

$$Complexity(s) = \frac{Service\_Granularity}{number\_of\_services}$$

$$= \frac{\sum_{i=1}^{|O(s)|} (SG(i))^2}{|S|}$$

where

- $|S|$—number of services to realize the application
- $SG(i)$—granularity of *ith* service.

Degree of control of a service calculated by the difference of SLC and DEP.

$$SLC = sum\_of\_CRUD\_coefficients\_for\_each\_Business\_entity$$

$$= \frac{1}{h_2 - l_2 + 1} \sum_{i=l_2}^{h_2} \sum_{sr \in SR} (BE_{i,sr} X V_{sr})$$

$$DEP = dependency\_of\_service\_on\_other\_service\_business\_entities$$

$$= \frac{\sum_{j=L1_i}^{h1_i} \sum_{k=1}^{BE} V_{sr_{jk}} - \sum_{j=L1_i}^{h1_i} \sum_{k=L2_i}^{j2_i} V_{sr_{jk}}}{nc}$$

$$Autonomy = \{SLC - DEP, ifSLC > DEP0\, otherwise,$$

where

- nc—number of relations with other services
- $BE_{i,sr} = 1$, if the service performs action sr on the ith BE and sr represents any CRUD operation and V_{sr} is the coefficient depending upon the type of action
- $V_{sr_{jk}}$—corresponding value of the action in $jkth$ element of the CRUD matrix gives the weight of corresponding business capability affecting a business entity
- $l1_i, h1_i, l2_i, h2_i$—bounding indices in the CRUD matrix of *ith* service.

Reusability is the number of existing consumers of the service.

$$Reusability = number\_of\_existing\_consumers$$

$$= |S_{consumers}|$$

Also, this can be evaluated from coupling, cohesion, granularity and consumability of a service, where consumability is the chance of the service being discovered and depends upon a fraction of operations in the service.

$$Reusability = \frac{Cohesion - granularity + Consumability - coupling}{2}$$

2.4 Microservice Development

A microservice architecture can be implemented as an enterprise application or a cloud service. In reality, it is identified as a cloud-based application using service-based architecture consisting of API-driven communication, polyglot service implementation, a set of loosely coupled services, service discovery capabilities, inbuilt failure tolerant technologies, deployed on containers and biased toward DevOps practices, which entails development and operations collaboration via automated service pipelines and agile development.

2.4.1 Microservice Design Principles

The design principles are based on the following principles:

- Every microservice should be managed, upgraded, scaled, and deployed independently
- Each microservice should have a bounded context and be specialized in the single functionality it delivers as a product on the process level
- Every microservice should be stateless, where possible, and designed for failure and recovery automatically
- For state management they should use existing caches, databases and directories.

The described drivers above should result in the following:

- Loose coupling
- Autonomy
- Reusability
- Composability
- Fault tolerance
- Service Discover capabilities
- Business process alignment via API.

2.4.2 Building Blocks

Microservice-based systems will be developed using design patterns and architectural styles, not bound to any specific technology limitations and modular in nature. They will communicate with other services via a lightweight data exchange mechanism such as REST.

Every microservice should provide a clear business use case.

2.4.3 Microservices Interaction

In monolith application development, each business component invokes procedure or function using a language-level call, such as function or method. In the microservice style, each service is self-contained, running an isolated process on its own distinct network node, and the communication will happen via an IPC (interprocess communication) mechanism.

There are lots of different IPC technologies to choose from. Services can use synchronous request/response-based communication mechanisms, such as HTTP-based REST or gRPC. Alternatively, they can use asynchronous, message-based communication mechanisms such as AMQP or STOMP. There are also a variety of message formats. Services can use human-readable, text-based formats such as JSON or XML. Alternatively, they could use a more efficient binary format such as Avro or Protocol Buffers.

Interaction Styles

There are a variety of client-service interaction styles. they can be categorized in two dimensions.

The first dimension is whether the interaction is one-to-one or one-to-many:

- One-to-one—Each client request is processed by exactly one service.
- One-to-many—Each request is processed by multiple services.

The second dimension is whether the interaction is synchronous or asynchronous:

- Synchronous: The client expects a timely response from the service and might even block while it waits.
- Asynchronous: The client does not, is not block, and the response, if any, isn't necessarily sent immediately.

Communicating Using the Synchronous Remote Procedure Invocation Pattern

When using a remote procedure invocation-based IPC mechanism, a client sends a request to a service, and the service processes the request and sends back a response. Some clients may block waiting for a response, and others might have a reactive, non-blocking architecture. Unlike messaging, the client assumes that the response will arrive timely.

Figure 2.3 shows how RPI works. The business logic in the client invokes a proxy interface, implemented by an RPI proxy adapter class. The RPI proxy makes a request to the service. The request is handled by an RPI server adapter class, which invokes

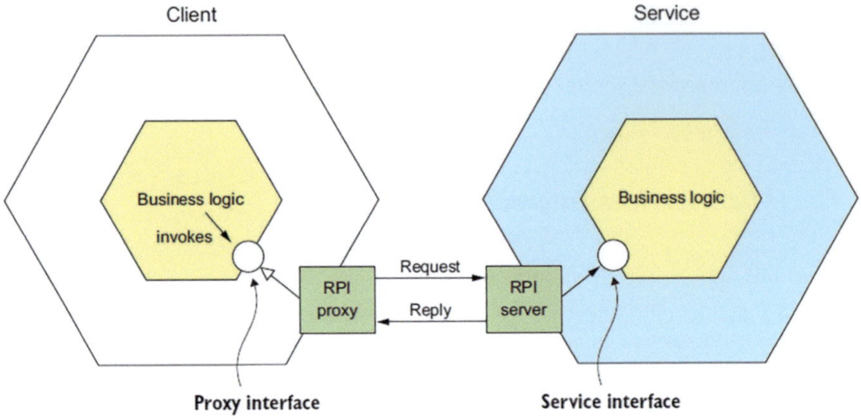

Fig. 2.3 Working of RPI

the service business logic via an interface. It then sends a reply to the RPI proxy, which returns the result to the client's business logic.

Using Rest

A key concept within REST is a resource, which typically represents a single business object, such as a Customer or Product, or a collection of business objects. REST uses the HTTP verbs for manipulating resources, which are referenced using a URL. For example, a GET request returns the representation of a resource, which is often in the form of an XML document or JSON object, although other formats such as binary can be used. A POST request creates a new resource, and a PUT request updates a resource. The Order Service, for example, has a POST/orders endpoint for creating an Order, and a GET/orders/{orderId} endpoint for retrieving an Order.

Rest Maturity Model

Leonard Richardson [4] defines maturity model for REST that consists of the following levels:

- **Level 0**: Clients of a level 0 service invoke the service by making HTTP POST requests to its sole URL endpoint. Each request specifies the action to perform, the target of the action e.g., for example, the business object, and any parameters.
- **Level 1**: A level 1 service supports the idea of resources. To perform an action on a resource, a client makes a POST request that specifies the action to perform and any parameters.
- **Level 2**: A level 2 service uses HTTP verbs to perform actions: GET to retrieve, POST to create, and PUT to update. The request query parameters and body, if any, specify the action-specific. This enables services to use web infrastructure, such as caching for GET requests.

- **Level 3**: The design of a level 3 service is based on the HATEOAS (Hypertext As The Engine Of Application State) principle. The basic idea is that the representation of a resource returned by a GET request contains links for performing actions on that resource. For example, a client can cancel an order using a link in the representation returned by the GET request that retrieved the order. The benefits of HATEOAS include no longer having to hard-wire URLs into client code (see Table B.3 of Appendix B).

Using gRPC

One challenge with using REST is that because HTTP only provides a limited number of verbs, it is not always straightforward to design a REST API that supports multiple update operations. An IPC technology that avoids this issue is gRPC, a framework for writing cross-language clients and servers.

gRPC is a binary message-based protocol, and this forces to take an API-first approach to service design. Any gRPC API uses a Protocol Buffers-based IDL, which is Google's language-neutral mechanism for serializing structured data. The Protocol Buffer compiler generates client-side stubs and server-side skeletons.

The compiler can generate code for a variety of languages, including Java, C#, NodeJS, and Golang. Clients and servers exchange binary messages in the Protocol Buffers format using HTTP/2.

A gRPC API consists of one or more services and request/response message definitions. A service definition is analogous to a Java interface and is a collection of strongly typed methods. As well as supporting simple request/response RPC, gRPC supports the streaming of RPC. A server can reply with a stream of messages to the client. Alternatively, a client can send a stream of messages to the server.

In a microservice application system the service can be developed in any technology stack. This promotes the concept, of technical heterogeneity which can be developed in any programming language and can use any kind of data storage technologies like relational databases, non-relational databases, graph databases or object-oriented databases etc. Thus, the constituting microservice can be developed using different programming languages like (C#, Java, Python, JavaScript, etc.). Services will be deployed on different nodes and the communication will be carried between them using remote calls but can be slow due to network latency, which by requires a lightweight communication protocol (see Table B.4 of Appendix B).

Communicating by Asynchronous Messaging Pattern

When using messaging, services communicate by asynchronously exchanging messages. A messaging-based application typically uses a *message broker*, which acts as an intermediary between the services, although another option is to use a brokerless architecture, where the services communicate directly with each other. A service client makes a request to a service by sending it a message. If the service instance is expected to reply it will do so by sending a separate message back to the client. Because the communication is asynchronous, the client does not block waiting for a reply. Instead, the client is written assuming that the reply were not received immediately.

Messaging Overview

A useful model of messaging is defined by Hohpe and Woolf. In this model, messages are exchanged over message channels. A sender (an application or service) writes a message to a channel, and a receiver (an application or service) reads messages from a channel [3].

A message consists of a header and a message body. The *header* is a collection of name-value pairs, metadata that describes the data being sent. In addition to name-value pairs provided by the message's sender, the message header contains name-value pairs, such as a unique *message id* generated by either the sender or the messaging infrastructure, and an optional *return address*, which specifies the message channel that a reply should be written to. The message *body* is the data being sent, in either text or binary format.

There are several different kinds of message:

- Document: A generic message that contains only data. The receiver decides how to interpret it. The reply to a command is an example of a document message
- Command: A message that is the equivalent of an RPC request. It specifies the operation to invoke and its parameters
- Event: A message indicating that something notable has occurred in the sender. An event is often a domain event, which represents a state change of a domain object such as an Order, or a Customer.

Messages are exchanged over channels. The business logic in the sender invokes a *sending port* interface, which encapsulates the underlying communication mechanism. The *sending port* is implemented by a *message sender* adapter class, which sends a message to a receiver via a message channel. A *message channel* is an abstraction of the messaging infrastructure. A *message handler* adapter class in the receiver is invoked to handle the message. It invokes a *receiving port* interface implemented by the consumer's business logic. Any number of senders can send messages to a channel. Similarly, any number of receivers can receive messages from a channel.

There are two kinds of channel:

1. Point-To-Point: Its channel delivers a message to exactly one of the consumers that is reading from the channel. Services use point-to-point channels for the one-to-one interaction styles. For example, a command message is often sent over a point-to-point channel.
2. Publish-Subscribe: Its channel delivers each message to all the attached consumers. Services use publish-subscribe channels for the one-to-many interaction styles described earlier. For example, an event message is usually sent over a publish-subscribe channel.

2.4.4 Service Discovery

Suppose a code invokes a service that has a REST API. In order to make a request, the code needs to know the network location (IP address and port) of a service instance. In a traditional application running on physical hardware, the network locations of service instances are usually static. For example, the code could read the network locations from a configuration file that is occasionally updated. However, in a modern cloud-based microservices application, it is usually not that simple.

A client is not statically configured with the IP addresses of the services. Instead, an application must use a dynamic service discovery mechanism (see Fig. 2.4). Service discovery is conceptually quite simple: its key component is a service registry, which is a database of the network locations of an application's service instances. The service discovery mechanism updates the service registry when service instances start and stop. When a client invokes a service, the service discovery mechanism queries the service registry to obtain a list of available service instances and routes the request to one of them.

There are two main ways to implement service discovery:

1. The services and their clients interact directly with the service registry
2. The deployment infrastructure handles service discovery.

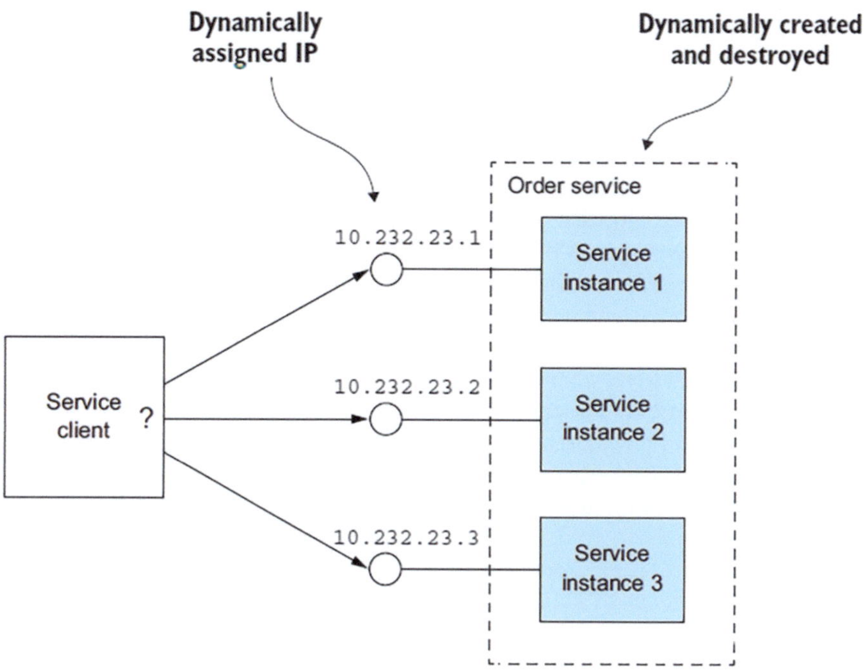

Fig. 2.4 Service instances having dynamically assigned IPs

Application-Level Service Discovery Patterns

One way to implement service discovery is for the application services and their clients to interact with the service registry. A service instance registers its network location with the service registry. A service client invokes a service by first querying the service registry to obtain a list of service instances. It then sends a request to one of those instances.

This approach to service discovery is a combination of two patterns. The first pattern is the self-registration pattern. A service instance invokes the service registry's registration API to register its network location.

Client-Side Discovery Pattern

When a service client wants to invoke a service, it queries the service registry to obtain a list of the service's instances. To improve performance, a client might cache the service instances. The service client then uses a load-balancing algorithm, such as a round-robin or random, to select a service instance. It then makes a request to a select service instance.

One benefit of application-level service discovery is that it handles the scenario when services are deployed on multiple deployment platforms. Imagine, for example, only some services deployed on Kubernetes, and the rest running in a legacy environment. Application-level service discovery using Eureka, for example, works across both environments, whereas Kubernetes-based service discovery only works within Kubernetes. One drawback of application-level service discovery is that it needs a service discovery library for every language.

Platform-Provided Service Discovery Patterns

Platforms such as Docker and Kubernetes have a built-in service registry and service discovery mechanism.

The deployment platform gives each service a DNS name, a virtual IP (VIP) address, and a DNS name that translates to the VIP address. A service client makes a request to the DNS name/VIP, and the deployment platform automatically routes the request to one of the available service instances. As a result, service registration, service discovery, and request routing are entirely handled by the deployment platform.

In a distributed system, whenever a service makes a synchronous request to another service, there is an ever-present risk of partial failure. Because the client and the service are separate processes, a service may not be able to respond in a timely way to a client's request. The service could be down because of a failure or for maintenance. Alternatively, the service might be overloaded and responding extremely slowly to requests.

Because the client is blocked, waiting for a response, the danger is that the failure could cascade to the client's clients, and so on, and cause an outage.

Circuit Breaker Pattern

This tracks the number of successful and failed requests, and if the error rate exceeds some threshold, actually trips the circuit breaker so that further attempts fail immediately. A large number of such failures means that the service is unavailable and that sending more requests is pointless. After a timeout period the client should try again, and, if successful, close the circuit breaker.

The "two pizza team rule" is a very dominating rule in the development of microservice-based applications. It is perceived as each microservice should be divided into a small team which can be fed by two pizzas. The idea is to handle and solve problems and reduce complexity.

Teams and group dynamics studies state that communications links between members grow with the team, following the equation:

$$\frac{(n * n - 1)}{2}$$

As the number of links grows within a team, their communication and productivity degrades.

Fluentd

- Data collector that gathers, unifies, filters, and routes log events.
- Works by tailing the container logs and gathers from both Stdout and Stderr.
- Tags log events with various metadata about the container.
- One instance for each node in the cluster.

ElasticSearch

- Responsible for storing all log events from Fluentd so they can be indexed and searched
- Each namespace can be stored in a different index or shipped to different clusters.

Kibana

- Tool for visualizing and analyzing the data stored in Elasticsearch.
- Installed by default in the cluster and configured to query the local Elasticsearch.

Prometheus is a monitoring system that stores time series data (see Fig. 2.5).

- Prometheus has service discovery capabilities with kubernetes
- Prometheus follows a pull model. when prometheus has an endpoint, whether static or discovered; it will go out and "scrape" the endpoint and pull the metrics via http
- Prometheus gathers these time series data and stores them with its efficient format on disk. This is not a distributed database
- with its storage format we are able to perform powerful queries that slice and dice for visualization and alerting to alert the users

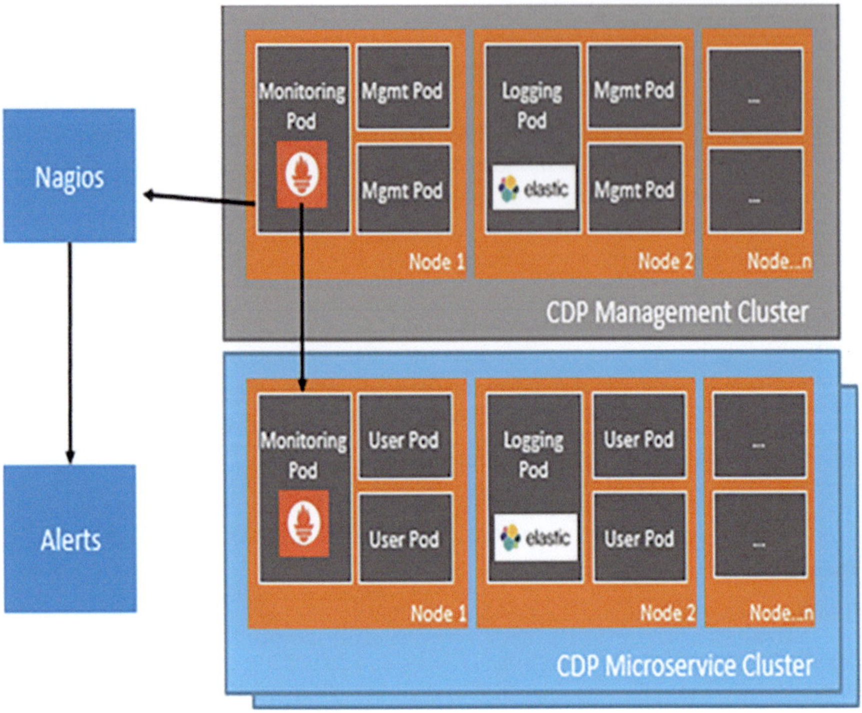

Fig. 2.5 Prometheus monitoring microservice stack

- Prometheus also has an exposed http api that clients can use to access metrics
- Prometheus will be able to aggregate metrics across clusters
- Integrates with Nagios for alerting.

Distributed Tracing is a process of collecting end-to-end transaction graphs in near real time. A trace represents the entire journey of a request. A span represents a single operation call. Distributed Tracing Systems are often used for this purpose. Zipkin is one example, as a request is flowing from one microservice to another, tracers add logic to create a unique Id.

Zipkin is a distributed tracing system implementation. Google Aggregate spans into trace trees and manages both collection and lookup of the data. It provides information such as Request Processing time, Request/Response for each span, Triggered URIs, etc.

2.5 Distributed Transactions in Microservices

When a monolith is converted into a microservice or developing a system via a microservice style, the transaction can span multiple services over the network that are called in sequence to complete the entire transaction.

Let us take an example of an e-commerce platform to understand it more deeply (see Fig. B.1 of Appendix B).

If a user sends a create order request to the platform, the platform will create a local database transaction, that works over various database tables, to execute the order and lock the items from the inventory. If at any point of time any step fails, the transaction will be rolled back to the point of create order and lock the item in the inventory.

However, refactoring this application into a microservice the order and inventory will decompose the a self-contained microservice which will be responsible for handling the create order transaction and the inventory as service, to lock the inventory item. Whenever a user tries to place an order the orchestrator is going to invoke both Order and Inventory Service (see Fig. B.2 of Appendix B).

Because the microservice is designed as a polyglot and has separate databases, it is a distributed transaction.

2.5.1 CAP Theorem

In 2000 Brewer said in his ACM conference keynote, "*You know, I will bet that a single distributed database can never realize the properties of consistency availability and partition tolerance all at once.*"

Let us define the following terms that make CAP theorem:

- Consistency
- Availability
- Partition tolerance.

CONSISTENCY is a property of a distributed system, where every read operation receives the most recent write or an error.

AVAILABILITY is the property of a distributed system where every request receives a (non-error) response, without any guarantee that it contains the most recent write.

PARTITION TOLERANCE is the property of a distributed system where the system continues to operate even if there is a partition/communication break.

In 2002, Gilbert and Lynch, published formal proof of Brewer's conjecture, rendering it a theorem (see Fig. 2.6): "*Of three properties of shared-data systems (Consistency, Availabilityand tolerance to network Partitions) only two can be achieved at any given moment in time.*"

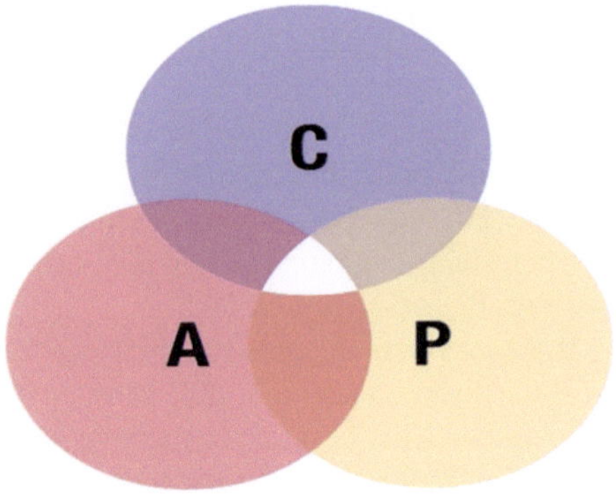

Fig. 2.6 CAP theorem

Here, CAP theorem states that every distributed system is on one side of the triangle, and there is no way to be in the center (see Fig. 2.7). Each side of the triangle has its own features, advantages, and disadvantages.

In cases of CA, figure as follows (see Fig. B.3 of Appendix B).

- The system achieves High Available Consistency.
- To manage data across multiple resources these systems used methods like two-phase commit
- Single-site databases, RDBMS (Oracle, Postgres, MySQL) could be good examples of CA systems.

In cases of CP figure as follows (see Fig. B.4 of Appendix B).

- The system achieves Strong Consistency with Partition Tolerance.

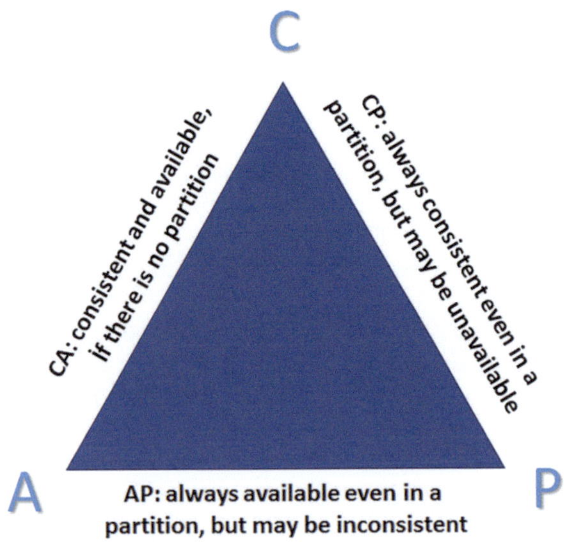

Fig. 2.7 CAP partition pyramid

- Pessimistic locking methods used for multi resource data management.
- DNS, MongoDB, Redis are examples of CP systems.

In cases of AP figure as follows (see Fig. B.5 of Appendix B).

- The system offers full availability by relaxing consistency.
- Optimistic locking methods used for multiresource data management.
- CouchBase, Cassandra, Hazelcast are examples of CP systems.

In monolith systems and databases having CA (Consistent and Available) capabilities, there is no distributed computing and no need to consider partitions.

Before distributed architectures and microservices were spread, only ACID transactions existed. However, after CAP theorem and increasing interest in distributed systems, achieving ACID became impossible. Brewer defined the BASE acronym.

ACID

The acronym ACID consists of 4 features:

- ATOMIC: "The transaction should either succeed or revert to the original state." In other words, all or nothing should be committed
- CONSISTENT: "Any transaction will bring the system from one valid state to another." Consistent here is different from consistent in CAP Theorem. The former is for the state of the whole system; however, the latter is about the consistency of a single entity
- ISOLATED: "Transactions cannot interfere with each other" This feature states that for a single entity, only one transaction can occur simultaneously
- DURABLE: "Once a transaction has been committed, it will remain so". This property is related to the persistence characteristics of the system. The database should remember everything after the transaction is completed.

BASE

BASE is an alternative multi resource data management methodology to ACID, described by Eric Brewer after proposing CAP Theorem. BASE consists of three properties.

- BASICALLY AVAILABLE: System mostly available but there might be some subsystems temporarily unavailable
- SOFT STATE: The state of a system could change over time, so even during times without any input being given there may be changes in the system taking place due to 'eventual consistency', thus the state of the system is always considered to be 'soft'
- EVENTUAL CONSISTENT: Given that if no new updates are made to a given data item, eventually all accesses will return the last updated value.

Consistency Types

- STRONG CONSISTENCY: Strong consistency is a consistency model where all subsequent accesses to a distributed system will always return the updated value after the update
- WEAK CONSISTENCY: It is a consistency model used in distributed computing where subsequent accesses might not always be returning the updated value. There might be inconsistent responses
- EVENTUAL CONSISTENCY: Eventual consistency is a special type of weak consistency method which informally guarantees that, if no new updates are made to a given data item, eventually all accesses to that item will return the last updated value.

2.5.2 Problems with Distributed Transactions in Microservices

Transaction Atomicity

In database systems, atomicity is known as that in any transaction either all steps will complete or no steps complete.

In microservices systems, by default, there is no global transaction coordinator.

Concurrent Request Isolation for User Actions

If the object is serialized by a transaction and, at the same time (before transaction completed), it is read by another request, should the object return the old data or updated data?

There are two patterns which are identified globally in the microservice development community:

1. Two-phase commit (2pc)
2. Saga.

The traditional approach to maintain data inconsistency across multiple services, databases, or message brokers, is by using distributed transactions. The industry recognized de facto standard for distributed transactions is the X/Open Distributed Transaction Processing (DTP) Model (X/Open XA), which uses two-phase commit (2pc) to insure all involved participants in a transaction, either commit or roll back.

An XA-compliant technology stack consists of XA-compliant databases and message brokers, database drivers, and messaging API's and interprocess communication mechanisms that propagate the XA global transaction ID. Most SQL databases are XA compliant and some message brokers also. In Java we can use JTA (Java Transaction API) to perform distributed transactions.

Problems with Two-Phase Commit

There are a variety of problems with distributed transactions, and are as follows:

1. Modern technologies like NoSQL, databases such as MongoDB and Cassandra, do not support 2pc.
2. 2pc is not supported by certain message brokers like RabbitMQ, Apache Kafka, etc.
3. 2pc is a form of synchronous inter process communication
4. All participating services should be present for 2pc.

Saga is a mechanism to maintain data consistency [3]. Saga is also perceived as a long-lived transaction (LLT) as it can hold on to database resources for a relatively long period of time, significantly delaying the termination of shorter and more common transactions. An LLT, has a long duration compared to the majority of other transactions because it can span multiple microservices and can also have lengthy computations to make.

Examples of LLT:

- A bank transaction where one account holds multiple services, so in this case, to fetch data from multiple bank services the transaction can take a while to produce results because it is accessing different services over this period
- An e-commerce platform which is decoupled these days, which will call different services to gather data as per the request type.

In most cases, LLT present serious performance issues as they are transactions the system must execute as an atomic transaction.

However, for specific applications it may be possible to alleviate the problems by relaxing the requirement that an LLT be executed as an atomic action. It should not sacrifice the consistency of the database, and it may be possible for certain LLTs to release their resources before they complete, thus permitting other waiting transactions to proceed.

Let us take an advanced approach with an example to illustrate the idea. In the example below, the hotel, car and flight booking might be done by different remote services. So, there is no technical transaction but a business transaction. When the flight booking cannot be carried out successfully, we need to cancel the hotel and car bookings (see Fig. 2.8).

2.5.3 Microservice Orchestration via Camunda BPMN

Saga is a long-lived transaction (LLT) flow. A sound approach to tackle this kind of transaction is BPMN (Business Process Model and Notation), which can directly execute compensation transactions for every failed transaction.

The model can be directly executed by a corresponding engine (see Fig. 2.9). The compensating activities are connected to the original activity. We can conclude here that a workflow engine capable of BPMN can act as Saga coordinator.

The example illustrates the code for adapter classes and compensating adapter classes in a process engine:

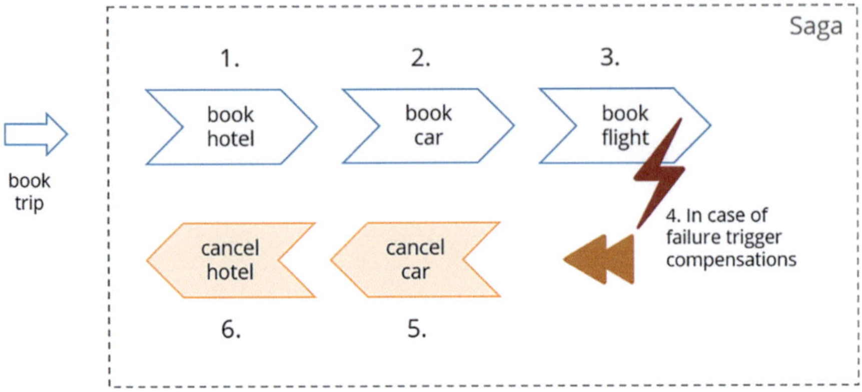

Fig. 2.8 BPMN hotel booking flow

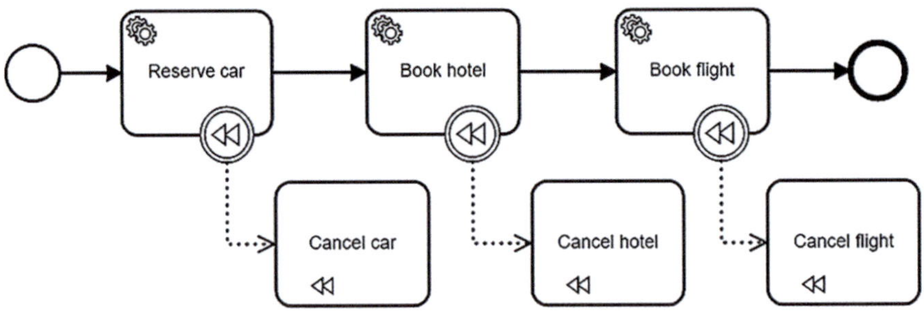

Fig. 2.9 Camunda modeler hotel book flow

```
SagaBuilder saga = SagaBuilder.newSaga("trip")
   .activity("Reserve car", ReserveCarAdapter.class)
   .compensationActivity("Cancel car", CancelCarAdapter.class)
   .activity("Book hotel", BookHotelAdapter.class)
   .compensationActivity("Cancel hotel", CancelHotelAdapter.class)
   .activity("Book flight", BookFlightAdapter.class)
   .compensationActivity("Cancel flight", CancelFlightAdapter.class)
   .end()
   .triggerCompensationOnAnyError();

camunda.getRepositoryService().createDeployment()
   .addModelInstance(saga.getModel())
   .deploy();
```

The solution described in the last section is through a process manager for the Saga pattern. The alternate way is a choreography approach which is perceived as routing slip.

To be sure that all are valid implementations for the Saga pattern, we refer to Rotem-Gal-Oz [22]. He states that *"I see sagas as the notion of getting distributed agreement of a process with reduced guarantees[...]—So, basically, a Saga is a loose transaction-like flow where, in case of failures, involved servicesperform compensation steps[...]".*

The definition using both a centrally managed process and a choreography process is further described as Saga. Therefore, for choosing the pattern it solely depends upon the requirements, as our usually preference is light weight workflow engines.

Saga is a very important topic to consider while developing microservice architecture for enterprise applications. The example described using Camunda orchestration for a microservice is mature and used widely.

2.6 Microservice DevOps/DevSecOps Automation

When organizations suffer data breach, companies do not only incur the cost of data damage and destruction, stolen money, business disruption and reputational harm. Other costs such as drops in share price, interruption to e-commerce, loss of customers and competitive advantage can also impact organizations affected by cybercrime. Often, security is an afterthought in SDLC. This section addresses the early adaptation of security in DevOps and DevOps practices for reliable and secure software systems development.

2.6.1 Operation and Security in IT Systems

Recently data breaches in the US include Target (data from up to 40 million credit and debit cards stolen in 2013), Anthem (80 million patient and employee records hacked in 2015), and Ashley Madison (more than 30 million user accounts hacked and released to the public in 2015). UK incidents include Sports Direct (employee data compromised in early 2017), Three Mobile (over 76,000 accounts hacked in 2016), Tesco Bank, and telecom company Talk Talk, which suffered a data breach in 2015 that cost the company a total of £80 million in fines. Australia led the Asia Pacific region in reported data breaches in 2016. The breaches referred to here, and the global WannaCry ransomware (May 2017), which utilized a well-known unpatched system flaw, demonstrate the consequences of ignoring security as a business priority. The organizations more rapidly developing, maintaining, and updating their applications have put security teams in a position of either slowing down release timetables or seeing their processes and controls circumvented.

As the volume of inputs needed to build applications, services, images, infrastructure as code and operations as code continues to expand, security teams will need a better way to gain visibility and control of these inputs. For example, are the sources of these inputs secure and trusted? And, how can security teams document the many system package dependencies needed for each? This lack of visibility and control handicaps their ability to secure the applications. They must walk a fine line between risk and reward on how to offer developers the freedom and flexibility they want, while still ensuring that the organization security posture robust and secure. An important goal for security in relation to DevSecOps is to ensure that only trusted code with the right privileges is used, and that vulnerable code does not get shipped. In this new frontier, where malicious actors frequently try to inject corrupt code into trusted public repositories, their job has never been more difficult. How can they take advantage of the potential of component modularity to accelerate business objectives and project timelines, while still ensuring secure code is being shipped?

DevSecOps is an approach aimed at directly bringing security philosophies and practices into the software development and release process through collaboration and automation [1]. It allows security teams to be both reactive (taking action on security issues after they are discovered) and preventive i.e., identifying, correcting and collaborating on security issues throughout the application lifecycle DevSecOps promotes automation of Infrastructure as Code (IAC) best practices throughout the process. It starts from initial application design and component selection, carries through to packaging and deployment, and encompasses post deployment monitoring and application maintenance, most notably security patching. This approach allows security teams to keep pace with development and operations teams and "cross pollinate" security best practices and innovations into existing development, build and release workflows. Security teams need to offer frictionless build and release guardrails to ensure that the components used by developers and operations teams are secure and releases are shipped and updated with the appropriate production grade configurations. Security teams that have modernized their cultural values and their technology stacks to accommodate this trend have reemerged as organization heroes, rather than the naysayers they have traditionally been seen as.

The traditional security approach is heavily focused on technology and its use by operations teams. It is typically a heavy-handed enforcement of policies laid down by security teams and implemented at or near the end of the development cycle. This is not an agile or iterative approach and it is unsuited to the fast release cycles of modern, cloud-centric IT operations. Shifting security left and making it a collaborative effort between development, operations and security teams, is the only practical approach. The traditional security approach is not agile or iterative and it is unsuited to the fast release cycles of modern, cloud-centric IT operations. Collaboration is the first step; automation is the next. According to the Oracle and KPMG 2019 Cloud Threat Report, only one in ten organizations are able to process over 75% of their security event data. As the number of applications, tools, processes, and deployment platforms that security teams must deal with on a daily basis rises, humans simply cannot keep up.

2.6.2 Security as an Element of Organizational Culture

Development teams focus on building or customizing applications. Operations teams focus on deploying and updating applications. DevOps allows development and operations teams to collaborate, but security teams are still disconnected from the process.

Fig. 2.10 Development and operations teams

Security is needed at every phase of the application development process, but when security teams operate in a separate silo, both development and operations teams suffer. Developers are increasingly drawing on external sources, such as snippets of code in GitHub or a container from a public repository, but many of them lack the detailed security knowledge needed to evaluate and code against vulnerabilities in these external sources. This can result in unsecure or malicious code making its way into production pipelines. In addition, many operations teams have inadequate experience in securing automation and deployment infrastructure, leading to a higher risk of compromise, security incidents and data leakage.

Making security an integral part of the DevOps toolchain allows development, operations, and security teams to break down traditional silos and have productive conversations about application, security, and operational best practices (see Fig. 2.10). With this approach, security teams have the opportunity to codify their training and knowledge into automated tools that development and operations teams can leverage throughout the application development lifecycle. This change does not happen with the implementation of a new tool or automation, it requires a culture change to take effect.

When security personnel are embedded into a collaborative organizational culture, development and operations teams get the help they need to deliver their desired outcome: secure applications running on secure platforms. While security teams have the confidence of knowing that processes are being enforced along the way, they have complete visibility into the provenance of component inputs, and what gets deployed is deployed in the most secure way possible (see Fig. 2.11).

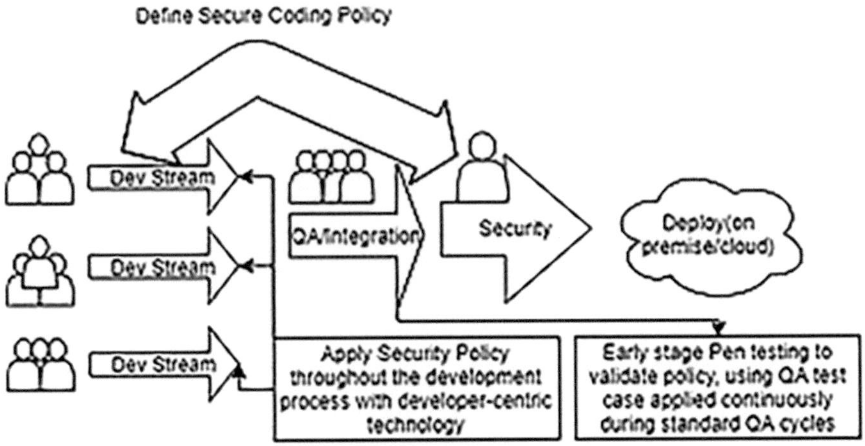

Fig. 2.11 Security team integration into DevSecOps

2.6.3 Frictionless Security Compliance at DevOps Speed

Automation is one of the key principles of DevOps. By automating different stages of the build release test deploy process, DevOps teams are able to quickly and reliably transfer applications from development to production. Security teams are typically unable to match this speed and, in the race to market, 52% of enterprises choose to sacrifice security.

In order to do their job effectively, security teams require greater visibility and control over the application delivery pipeline. For example, every application should have a complete and detailed list of all the system packages, tooling and third-party components that go into it, including version numbers and other metadata. Similarly, every application should have an immutable deployment template, which packages all the application artifacts and the automation needed to deploy them, and also configures all necessary logging, intrusion detection and security monitoring tooling. However, a recent survey found that only 38% of enterprises have a complete software "bill of materials" for each application.

This lack of visibility affects the ability for security teams to create automated, frictionless security checks and provide the security assurances the business expects, not just for new applications, but for every application released or used by the company. At the same time, the diversity of new attack vectors and the speed at which patches are introduced can quickly overwhelm a security team capacity to keep pace, and may result in a development crisis.

Continuous security testing offers a solution to this problem. There are two practices for this approach: creating security tests for attack vectors (similar to unit and functional tests), and continuously scanning application artifacts and checking dependencies, system packages and third-party software against trusted security databases for updates and Critical Vulnerabilities and Exploits (CVEs) that may affect them.

A failure in the former prompts development teams to write more secure code; a failure in the latter prompts operations teams to initiate a rebuild of the application components, repackage them with their service dependencies and redeploy them.

Taken together, these two make a pattern to reduce the potential surface area for attack and improve the organization's overall security posture. When these security hardened build artifacts are packaged using deployment best practices and the security tests are integrated with the DevOps pipeline, the "speed versus security" debate becomes irrelevant because security checks can be conducted continuously, as needed, and with the same speed and thoroughness as build and release checks.

2.6.4 Programmatic Enforcement of Trust

Trust is a foundational element of DevOps culture, and ensuring trust is a key goal of the DevSecOps culture. Current image and deployment template generation techniques typically rely on individuals or teams to generate and package artifacts for deployment. When it comes to security, there is often a disproportionate reliance on the human element. DevOps teams attempt to enforce trust by using multiple automated tools and services for artifact storage, application packaging, CI/CD, testing, and release management. However, a significant element of adding, configuring, auditing and updating security tooling in image artifacts and deployment templates is manual, relying on the training and knowledge of security professionals. This is a major gap for DevOps pipelines because it places a huge burden on individual performance and fails to account for inadvertent misconfiguration, introduction of local environment variables or vulnerabilities, actors not following defined processes, or malicious internal or external actors.

Enforcing the use of trusted sources, such as certified public repositories (for example, RedHat for the operating system and Bitnami for open-source software applications) or private enterprise repositories, carries the least risk for the organization, but also reduces development flexibility due to the limited components available. At the other end of the spectrum, relying on open-source resources and the "wild west" of the public GitHub repositories enables access to the latest, bleeding edge components and features, but comes with a corresponding loss of security oversight and control. Another problem is upstream dependencies, many of which may not be fully known to security teams and therefore lie beyond their boundaries of inspection. If any upstream dependencies are compromised but go undetected, they can wreak havoc with an organization's DevOps pipeline, leaving applications unsecured against the latest CVE, or breaking release automation due to compatibility issues in a package update. Dependency chains can be very long, with multiple embedded elements including low-level libraries that developers might use. Even when security teams are well-informed and aware, the speed at which new releases are made and the increasing complexity of these dependency trees can quickly overwhelm their ability to audit and enforce security rules. Finally, security policies and requirements are far from static. Teams have been largely trained to think of security as a point-in-time exercise—for example, releasing a new operating system image every 30 days. However, in the continuous battle with nefarious actors, and due to the ever-increasing list of CVE in fundamental packages and libraries, this approach is no longer tenable. Security demands a process that is as much, if not more, of an iterative process than application development. More importantly, security teams need to be able to both identify events and then propagate updates and patches through complex systems as quickly as possible.

One solution is to generate artifacts and enforce security policies programmatically. For example:

- Fetching inputs (code and scripts) from defined CI, artifact repository and source control tools on specific, locked-down networks or specific trusted and certified sources, such as a known operating system vendor, cloud provider, or secure open-source application provider, authenticated by a service account
- Generating images and deployment templates within public or private cloud accounts that are controlled by the enterprise and/or reside on trusted networks
- Automatically enforcing, through automation, the application of a security hardening configuration to every single artifact created, as a matter of process, and ensuring that this process can be updated seamlessly and independently from developer work
- Delivering generated outputs to only secure enterprise-managed repositories, container registries or cloud storage accounts.

"When trust is programmatically enforced at every stage of the DevOps pipeline, development, security, and operations teams have the confidence that what is eventually deployed is sourced from only secure and approved repositories". A team that configures its process and tools to point to a centralized packaging pipeline that generates its required artifacts, no longer has to think about custom built steps in its tools, nor does it need to be interrupted to update those custom build steps every time security needs change. It can simply trust that the needed security configurations and best practices are automatically factored in. Similarly, security teams can react to events and regenerate patched artifacts independently of a development team's process, then send those downstream to operations for deployment.

"When trust is programmatically enforced at every stage of the DevOps pipeline, development, security and operations teams have the confidence that what is eventually deployed is sourced from only secure and approved repositories." Security teams attempt to solve this problem by generating and maintaining custom "golden images", baseline templates that include all necessary enterprise security policies and that must be used by other teams. This is a partial solution but comes with its own challenges related to artifact maintenance, distribution, and enforcement of usage. It is important to consider this process as a continuous one and realize the interrupt effect it has on teams that must adopt and consume this image and refactor their applications on top of it independently. Simply making an artifact available does not mean it is going to be used, and the redeployment of applications on these new foundations is the most critical step of the process.

A better solution is to treat security policies as code and integrate the hardening process into the existing DevOps pipeline. This ensures that the final images and deployment templates, no matter the environment, automatically have all the necessary hardening and security configuration required by the enterprise without any additional overhead on development and operational teams. This approach also allows the testing framework, built into existing DevOps toolchains, to be extended to security, enabling security teams to independently update security policies and configurations on the fly without interrupting development.

Implementing security policies as code enables enterprises to transition to a layered packaging approach. Typically, this would involve starting with a minimalistic off-the-shelf operating system base layer containing only a limited set of necessary libraries and packages, then securing this foundation with a second layer of additional enterprise configuration and security policies. With this secure foundation established, additional layers would be added for specific components, application runtimes, dependencies, and code from custom enterprise repositories. Configuration automation would propagate any changes in base layers to dependent layers.

2.6.5 CLD Approach

The diagram representing this approach (see Fig. B.6 of Appendix B) includes:

- Variables (Market Size, Potential Customers, People Buying Product, etc.)
- Connections: S (same)—one variable is directly proportional to the second (for example, the more people who bought the product, the larger the customer base); O (opposite)—one variable is inversely proportional to the second (for example, the more people who buy the product, the less potential customers remain)
- Symbols R and B are usually designated cycles or feedback loops: R (reinforcing loop)—reinforcing cycles; B (balancing loop)—balancing loops.

"A downward spiral caused by a conflict of interests between development and operation, leads to a slower time for outputting solutions to a product, lowering the quality of services, increasing the number and duration of failures, accumulating problems and permanently extinguishing fires."

The nature of the phenomena above is due, as we have long known, to two conflicting goals of DevSecOps, (see Fig. B.7 of Appendix B):

- Support business development and make fast changes
- Provide high quality services (e.g., security, reliability, usability).

The faster are the changes (less Time to market), the is the greater risk in terms of service security (less Service Security). Of course, Time to market and Service Security affect each other not directly, but through related variables. Time to Market (Lead Time) is made up of Process Time (time to work on the change) and Queue Time (time to wait for the change in the queue). In turn, the quality of security

(Service Security) is inversely related to the risks of changes (Change Risk) (see Fig. B.8 of Appendix B).

Some of the key variables that help to balance Agility vs Stability are:

- Release rate—frequency of implementations.
- Release size—the average implementation size (lines of code, the number of release units in one release etc.).

Frequent introduction (higher Release rate) to help re duce the size of the queue (less Backlog Size). In this case, a long line of changes, very likely, will push us to "amalgamate" implementation (to improve Release Size), which again increases the risks (Security Risk) (see Fig. B.9 of Appendix B).

In turn, the higher the risk, the greater the number of implementation fails, which require additional changes to address the consequences or results of adjustments and result in Backlog size growth.

This is important as the dependence is described as a reinforcing feedback loop. Without deliberate action on the system the growth of any of the variables, this will lead to its further growth.

On the other hand, the growth of Backlog Size leads to changes in latency queue and, consequently, of Time to Market.

Process Time is largely determined by our ability to make changes (Change capability), which is defined by:

- Resource base available for changes.
- Level of IT professional competencies.
- Availability of knowledge and experience, ensuring free flow of information (see Fig. B.10 of Appendix B).

One of the important messages in DevSecOps is: The higher the Release Rate, more often are lessons learn (see Fig. B.11 of Appendix B).Thus, we can observe another reinforcing feedback loop, this time a positive one.

Process Time also depends on how smoothly the changes are made—how many changes are implemented the first time without returning to previous stages or finalizing (First-Time Implementation Rate), (see Fig. B.12 of Appendix B).

A high Change Control Level reduces Change Risk, but at the same time takes longer (higher than Process Time), and this is a major fork in change management (see Fig. B.13 of Appendix B).

Of course, this is not an exhaustive set of factors influencing Time to Market and Service Quality, but the diagram includes most of the key variables related to DevSecOps, and on this basis, it seems, it is already possible to illustrate important phenomena of the IT management system (see Fig. B.14 of Appendix B).

2.6.6 *Failures and Workarounds*

It is associated with a systematic increase in the complexity of the infrastructure, which is also poorly documented and is, due to the huge number of unsolved problems and workarounds, extremely fragile (see Fig. B.15 of Appendix B).

The more unsolved problems and workarounds, the higher the IT Infrastructure Complexity. On the one hand, this increases the fragility of systems which results in greater "readiness" even from minor changes (higher than Change Risk), and on the other hand complicates planning (reduces Change Control Level). In addition, there is time to document, which means less and less knowledge to control changes, and increasing dependency on specific people, saviors. Increased risks lead to new failures and new workarounds, which immediately form two reinforcing feedback loops.

Best Practices

Successful security programs involve three intersecting parts: people, processes, and technologies. DevSecOps is no different, but DevSecOps recognizes that security is the responsibility of everyone in an organization, and everyone has a role to play.

People

People are the starting point of DevSecOps implementation. Through ensuring proper training and restructuring of teams, security becomes a frame of mind rather than a hindrance. No matter there are many technologies, the weakest link will always be the human factor. This is the starting point for any DevSecOps implementation. One of the most important, but equally most difficult, aspects of DevSecOps is challenging the way traditional security teams integrate with the wider business. Most people approach risk from a place of denial rather than acceptance and preparation. Changing habits and raising awareness across all levels of a company are not easy tasks and require a top-down approach if attitudes are to change. Security needs to shift from being exclusive to being inclusive to facilitate this culture change. By integrating security teams within development teams, as DevSecOps does, companies will obtain earlier feedback on the quality, from a security point of view, of the code, software, or application, thus reducing the costs associated with implementing these fixes. Hiring security specialists, giving them a voice in project delivery, and allowing them to integrate their processes in the agile development world, will deliver the necessary results. Agile development helps to speed up product release dates, but often at the cost of neglecting security.

Process

DevSecOps aims to align and implement processes common to an enterprise to facilitate cooperation and achieve more secure development processes as a whole. People implement the processes in any organization, although processes have been siloed per team and are often not interconnected with other teams, and are, thus, counterproductive within an enterprise. DevSecOps aims to align and implement processes, common to an enterprise, to facilitate cooperation and achieve more secure development processes as a whole.

The Key points for essential processes of DevSeOps:

- Version control
- Integration Process
- Security Tooling CI/CD
- Compliance
- Security Architecture
- Incident Management
- Red Teams & Bug Bounty
- Threat Intelligence.

Processes are key to the success of DevSecOps. Their aim is to create agreed and repeatable ways of working which are clearly documented and available to the company to ensure transparency of the security protocols. If these agreed ways of maintaining security (sets of principles, playbooks, as defined above) are implemented, problems (faults, bugs, threats etc.) can be automatically identified much sooner and responded to in an agile fashion.

Technology

Technologies enable people to execute DevSecOps processes, which aim to reduce the enterprise attack surface and enable effective management of the technical security debt. Technologies are what enable people to properly execute DevSecOps processes.

The technologies required to implement a successful DevSecOps methodology within the enterprise include:

- Automation and configuration management
- Secure coding practices/Security as a code
- CI/CD for patching
- Application-level Auditing and Scanning
- Automated Vulnerability Management Scanning
- Automated Compliance Scan
- Managing Secrets.

Technologies are key, and the successful implementation of the ones listed above will greatly reduce the enterprise attack surface as well as the ability to effectively manage their technical security debt. As well as automating scanning and monitoring, implementing the above technology practices means that security, regulatory and compliance requirements, can be embedded as code into the software delivery pipeline to ensure that any code deployed is secure and compliant. Any deviation from this can be spotted early and fixed quickly. Ownership of these technologies does not need to reside within the security team and properly distributing these to the relevant operational team will help the security team to concentrate on hunting the threat rather than operating the known technologies.

DevSecOps addresses the need for proactive, customer-focused security that anticipates rather than reacts to data breaches or other cyberattacks. DevSecOps shifts security from reactive to proactive, supported by different techniques such as Test-Driven Development and Attack Driven Defense. It champions the importance of security at all levels of an organization and empowers security staff to make decisions that have a positive influence on their business. As such, DevSecOps, as both a concept and a practice, is growing all the time, with an increasing number of organizations implementing DevSecOps as a solution to their security issues. The benefits DevSecOps brings to companies that embrace it are numerous, including cost reduction, speed of delivery, speed of recovery, compliance at scale, and threat hunting. The cumulative effect of these benefits is an enhanced business reputation and a smoother business model. A DevSecOps enterprise will have successfully removed the barriers between DevOps and Security, helping them to work as one toward the enterprise business goals without any friction. The DevSecOps enterprise will have the ability to detect and fix security issues earlier in the development process, thus greatly reducing the cost associated with identifying and fixing them. Shifting security to the left through the use of people, processes, and technology will help to achieve this goal.

2.7 Conclusion

This Chapter focused on microservice-based architectures and DevOps-related techniques; as a result, it identified patterns and practices to integrate development and operation in sustainable and crisis-resilient digital production.

References

1. Swartout, P.: Continuous delivery and DevOps—a quickstart guide: Start your journey to successful adoption of CD and DevOps, 3rd edn. Packt Publishing, (2018). ISBN 978–1788999083
2. Agrawal, D., Das, S., El Abbadi, A.: Data management in the cloud: Challenges and opportunities. Springer, (2012). ISBN 9781608459254.
3. Kleppmann, M.: Designing data-intensive applications: The big ideas behind reliable, scalable, and maintainable systems. O'Reilly Media, (2017). ISBN 978–1449373320.
4. Richardson, C.: Microservices patterns: With examples in Java. Manning, (2018). ISBN 978–1617294549
5. Newman, S.: Building microservices: Designing fine-grained systems, 2nd edn. O'Reilly Media. ASIN B09B5L4NVT, (2021)

Chapter 3
Going Smart: Processes Re-Engineering for Hydraulic Fracturing

Abstract This chapter provides a case study of process-based transformation for an oil-and-gas enterprise related to hydraulic fracturing. This change is typical for growing, volatile, and competitive backgrounds with complex business landscapes that have the potential to trigger crises. Similar to the examples discussed in the previous chapters, this critical step toward digitalization requires cutting edge techniques and tools to gain success. Obviously, the oil-and-gas business is IT intensive as it integrates multiple system levels, such as SCADA, MES, and ERP. In this case, the integration followed a data-centric pattern supported by a dashboard tool that provided the top management with a smart BI-based graphical interface. This intelligent tool allowed for efficient data integration and instant multi-context scenario switching, which improved their decision-making processes and eventually fostered the business transformation.

Keywords Object-based optimization · Optimization methods · Tradeoff analysis

3.1 ANALYSIS of the Oil Industry and Prerequisites for Digitalization of Business Processes

3.1.1 Introduction

With the increase in the amount of data, the need for automated reporting is becoming more urgent and affects all areas of activity. To make efficient data-driven management decisions, managers need to track various metrics and assess the full picture. This problem is acute in the oil-and-gas industry, since for effective management it is necessary to quickly track various production indicators, for example, the planned and actual time of work, the amount of resources used, and the number of units of machinery and equipment used.

Currently, a large number of reports in oil-and-gas companies is not automated, which becomes a problem because the data is stored in disparate files, making it difficult for employees to share information in a timely manner, which in turn affects the

S. V. Zykov, *IT Crisisology Patterns and Practices*, Smart Innovation, Systems and Technologies 449, https://doi.org/10.1007/978-981-95-1260-7_3

decision-making. To speed up the receipt of information, it is necessary to automate the reporting process. There is a need to consolidate data in a single place, which will speed up access by staff and increase the effectiveness of their decision-making.

To solve this problem, it is necessary to pay attention to data visualization tools that quickly present information in a convenient and understandable form, in various sections for the desired time periods, and then analyze it to identify trends, patterns, and potential problem areas.

The purpose of this section is to develop analytical BI reporting for the management of oil production processes using hydraulic fracturing.

Objectives of the study:

- Analyze modern methods of oil production
- Analyze the main problems faced by companies in managing the oil production process and to propose possible approaches to solving the problem
- Analyze the market of Business Intelligence descriptive analytics tools;
- Identify the company's goals and key business needs
- Choose an appropriate methodology to collect requirements for a descriptive analytics tool
- Develop analytical reporting based on the selected Business Intelligence descriptive analytics tool
- Describe the effects after the implementation of the tool.

3.1.2 Overview of Oil Production Methods

The oil-and-gas industry is one of the main high-tech sectors of the Russian economy.

There are three main classes of oil production methods [6]:

- Primary Recovery is a method in which oil is extracted due to natural internal forces, such as reservoir elasticity, dissolved gas energy, gravity, surrounding waters and gas cap
- Secondary recovery includes methods of artificially maintaining reservoir pressure by injecting water or gas, which helps to conserve energy and increase oil recovery
- Tertiary recovery uses enhanced oil recovery (EOR) methods by injecting special agents into the reservoir that increase the displacement of oil from the reservoir.

EOR methods are aimed at extracting residual oil from reservoir zones that were poorly involved in the previous production process. These zones can be flooded, gassy, and poorly permeable with high oil saturation [5].

A key feature of EOR is their narrow focus, i.e., they affect only some of the factors that affect residual oil. There is no universal method equally effective in all cases due to the variety of residual reserves and properties of oil, water, and gas in different fields [5].

To achieve maximum EOR efficiency, it is necessary to take into account the individual characteristics of each field and carefully select the most suitable method from the existing ones [5].

EOR is classified according to the type of working agents used as follows [4]

(1) Thermal methods

- steam-thermal effect on the reservoir
- in situ combustion
- displacement of oil by hot water
- cyclic steam treatment of wells.

(2) Gas methods

- air injection into the reservoir
- impact on the reservoir by hydrocarbon gas
- the impact of carbon dioxide on the reservoir
- impact on the reservoir by nitrogen, flue gases, etc.

(3) Chemical methods

- displacement of oil by aqueous solutions
- displacement of oil by polymer solutions
- displacement of oil by alkaline solutions
- displacement of oil by acid solutions
- displacement of oil by chemical reagent compositions (micellar solutions, etc.)
- microbiological effect.

(4) Hydrodynamic methods:

- integrated technologies
- involvement of non-drainable reserves
- barrier flooding in gas and oil deposits
- non-stationary flooding
- forced fluid withdrawal; step-by-thermal flooding.

(5) Combined methods (mostly used)

- hydrodynamic and thermal methods
- hydrodynamic and physicochemical methods
- thermal and physicochemical methods.

(6) Physical methods of increasing the flow rate of wells differ from classical EOR in that they use the natural energy of the reservoir and do not increase the potential of the displacement agent. Also, physical methods do not affect the final oil recovery, but only temporarily increase the flow rate of the well, affecting only the current oil recovery. These methods include the following:

- hydraulic fracturing (hydraulic fracturing)
- horizontal wells

- electromagnetic impact
- wave action on the reservoir.

3.1.3 Description of Hydraulic Fracturing Technology

Much of oil reserves are classified as hard-to-recover because they are enclosed in reservoirs with low permeability and limited drainage [12]. Hydraulic fracturing (HF) is used to ensure the efficiency of extracting natural resources from such deposits. In the case of successful hydraulic fracturing, as a rule, oil production increases sharply at idle wells, where traditional methods no longer bring economic benefits or are difficult to implement. The right choice of a well plays a key role before oil production by hydraulic fracturing and prevents economic losses. Oil production by hydraulic fracturing has been carried out since the middle of the last century and still does not lose its relevance due to the depletion of easily accessible reserves and the predominance of a large share of hard-to-recover oil under thick rock layers [12].

During hydraulic fracturing, gel-like liquids are injected into oil wells under pressure of such strength that many fractures are formed in the rock formation, which subsequently become channels for oil inflow. If injection is then suspended, the fractures will gradually close under the pressure of the rocks, and the necessary conductivity effect for oil inflow will be lost. To avoid this, proppant is pumped into the cracks along with the liquid - a strong and resistant bulk material of a very small caliber, similar to sand. The liquid acts as a proppant transporter into the cracks and subsequently self-destructs. In turn, the proppant prevents the closure of the fracture and creates a drainage effect, due to which microscopic channels are fixed in the formation. Depending on the depth of the formation, different types of proppant are used, differing in composition, cost, and strength level (see Fig. 3.1).

This method has proven to be highly effective and cost-efficient and continues to improve in terms of its methodology, materials used, and specialized equipment and machinery [13].

3.1.4 Automation of Management Processes in the Oil-and-Gas Industry

In the context of the rapid development of the oil-and-gas industry, the market is becoming more complex, and the projects scale is growing. Effective management of processes and business decisions is critical for the successful implementation of projects [14].

To ensure operational management, it is necessary to use state-of-the-art methods and tools, such as:

- Business process management and automation systems

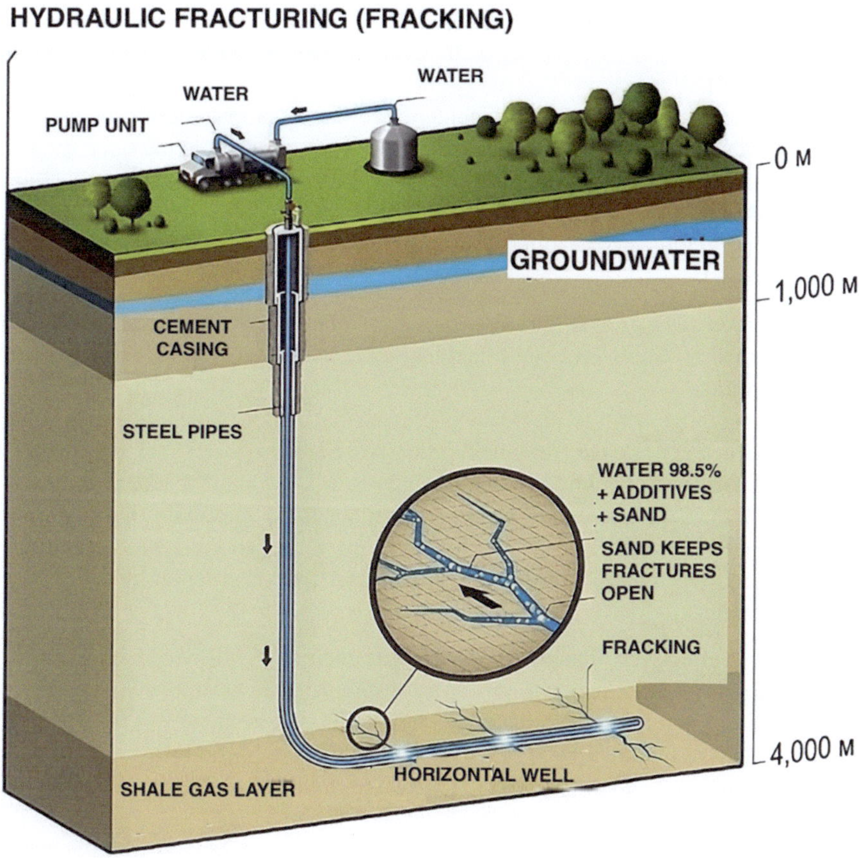

Fig. 3.1 Hydraulic fracturing [21]

- Multi-criteria models
- Risk analysis systems, etc.

Many of the technologies included in the Digital Economy of the Russian Federation project are already being implemented in the oil-and-gas industry [8]:

- big data
- neurotechnology and artificial intelligence
- distributed ledger systems
- quantum technologies
- new production technologies
- Industrial Internet
- components of robotics and sensors
- wireless communication technologies
- virtual and augmented reality technologies.

Process management in the oil-and-gas industry is a complex task due to the high level of risks and multiple legislative, environmental, technical, and financial constraints [14].

In the oil-and-gas industry, management decision-making requires an in-depth study of production activities, taking into account many factors. At the same time, of data collection from various systems often leads to manual reports generation, which not only takes a lot of time, but also creates the risk of errors.

According to the classification of Andersen, business processes are divided into three groups [7]:

- Main processes (production) are processes that create the main value for the enterprise. Such processes include: extraction, processing, storage, and transportation
- Supporting processes (auxiliary) are processes that do not create immediate value, but are necessary to ensure the main processes. Such processes include: financial management, personnel management, diagnostics, and maintenance and repair
- Developmental processes are processes that focus on improving the main and supporting processes, increasing their efficiency and creating new opportunities. Such processes include investment projects, reconstruction, and technical re-equipment

In order to effectively manage an oil-and-gas enterprise, it is necessary to ensure the integration of information flows from different systems into a single digital platform database for monitoring and managing all processes.

Safety and risk reduction at work are priority areas. Monitoring and control systems allow the authorities to quickly respond to emerging problems and situations, thereby minimizing the likelihood of accidents and accidents.

Personnel and equipment monitoring technologies allow for tracking the actual use of labor and equipment in projects. These systems monitor the efficiency of the time spent on production and ensure control of compliance with industrial and environmental safety standards.

Automation of business processes and consolidation of data into a single corporate system of oil-and-gas enterprises leads to acceleration of managerial decision-making, improved control over production, reduction of costs, and optimization of the resource use. Single integrated information and management system of the enterprise allows the analysis of production data and control and modeling of production processes, as well as effective planning [14].

3.1.5 Analysis of Analytical Levels and Their Application in Managerial Decision-Making

Oil production by hydraulic fracturing is a complex process that must be effectively managed to achieve the desired results. Digital transformation plays an important role in the management process.

The main problem is the low level of efficiency of economic process management [15]. To make effective management decisions, it is necessary to rely on data [16].

Experts distinguish four levels of analytics, each of which differs in the level of complexity of the information processing models used and the degree of human participation [1]:

- Descriptive
- Diagnostic
- Predictive
- Prescriptive.

Let us take a closer look at each of these levels (see Fig. 3.2).

Descriptive or descriptive analytics offers a retrospective view of events and provides an answer to the question, "What happened?" This type of analytics is the most common and is based on events that have occurred in the past. The collected data are useful in further analysis of indicators and helps to understand such things as [1]:

- What were the company's sales?

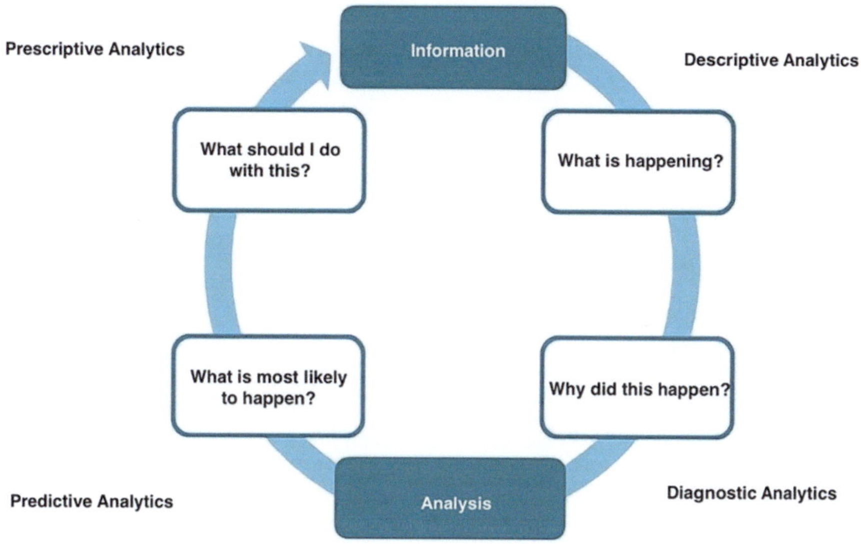

Fig. 3.2 Types of analytics

- How productive was the company as a whole?
- What was the outflow/inflow of customers during the period?

Descriptive analytics includes designing a data model, selecting appropriate visualizations, and building dashboards [17].

This type of analytics allows experts to quickly receive information about the state of the company and make management decisions for its further development. Analysis of historical data allows the identification of problem areas of activity. However, descriptive analytics is only the initial stage in data analysis, and other types of analytics need to be used to get the most out of it [1].

The diagnostic analytics analyzes data at a deeper level and identifies the relationships of events, non-obvious dependencies, patterns, and key influencing factors. Similar to the previous type of analytics, diagnostic analytics is based on historical data, but it serves to answer another question: "Why did this happen?" While this level is very important, it is often ignored because of the tendency to move from a basic analysis of "what happened" to a prediction of "what will happen" without paying due attention to the intermediate stage of "why it happened". Thanks to diagnostic analytics, companies can answer the following questions [1]:

- Why did the company's sales drop?
- Why did the company's productivity fall?
- Why was there an outflow of customers?

Diagnostic analytics is more complex due to the use of data mining, probability theory, and time series technologies. At this level, the anomalies in the data are identified, which allows companies to quickly detect and prevent manufacturing defects or fraudulent activity [2].

Descriptive and diagnostic analytics are based on the data available, thanks to which it is possible to analyze the results produced, which is already sufficient for making management decisions. However, in order to go beyond the analysis of historical events and be able to make informed forecasts about the future, it is recommended to use more advanced types of analytics.

Predictive analytics relies on the previously described types of analytics as a basis for building models and answers the question "What can happen?". Predictive analytics methods first identify and then interpret the identified patterns [3].

Predictive analytics solves the following tasks:

- Forecasting the company's performance
- Predicting maintenance problems and equipment breakdowns.

With the growth of data, predictive analytics become increasingly important in various industries. The unstructured nature of much of this data requires in-depth investigation through the use of predictive analytics. However, due to the difficulties in their processing of such data, many companies do not try to structure them [1]. The results of predictive analytics can be used to improve business processes, increase productivity and increase production efficiency.

The fourth level of analytics—prescriptive analytics—provides an answer to the question "What to do now?" and recommends specific actions, simplifying the decision-making process. Recommendations are based on the results of descriptive, diagnostic, and predictive analysis. Predictive analysis is aimed at determining the optimal solutions in the current situation or measures leading to the most favorable results. Historical data analyzed using descriptive and diagnostic analytics is used to create predictive forecasts and model the optimal company development. In addition, modeling the probability of various outcomes allows for making informed management decisions.

Creating prescriptive models requires more complex combinations of methods. Machine learning is gives way to artificial intelligence and specialized algorithms designed to solve specific problems [1].

3.1.6 *Prospects for the Implementation of BI Systems for Monitoring Economic Activity*

The concept of Business Intelligence (BI) is characterized as a set of tools that allows the automation of the transformation of raw data into valuable information for making management decisions. The process includes sequential data processing: automation of extraction, transformation, and loading (ETL) processes, data storage, data visualization, and user interaction [18].

Business intelligence systems perform the following functions [22]:

- Integrate data from various sources into a single system, providing access to up-to-date information in real time
- Reduce the time for creating reports
- Create a unified information environment in the organization eliminating data silos
- Simulate business situations within a single system
- Automate predictive analytics, predicting future trends
- Maintain stable operation when the volume of processed data increases
- Collectively create reporting
- Prompt data analysis at the request of users.

The advantages of implementing BI solutions include the following:

- Improving the quality of management decisions due to the ability to use structured information in a data warehouse, presented in a convenient form for quick visual analysis
- Fast of access to decision-making information, eliminating intermediate stages of data uploading and processing with the involvement of specialized programs and specialists
- View reports on a variety of devices, including computers, laptops, tablets, and smartphones, providing secure remote access. This allows remote workers to have

prompt access to analytics, which contributes to making informed decisions based on data
- Transparency of processes through visualization and monitoring of quantitative indicators
- Forecast based on historical data, which contributes to strategic planning
- Identifying bottlenecks, growth points through monitoring indicators.

BI brings benefits, including increased speed and accuracy of business analysis and decision-making, improved employee satisfaction and customer loyalty, increased revenue, and reduced costs [9].

In manufacturing companies, BI reports are used to analyze equipment operation and quality control, technical problems in the operation of equipment and the need for repairs, the capacity, and the volume of output and resources consumed. This contributes to improving the efficiency of production management and minimizes crisis consequences [10].

The integration of BI systems with machine learning technologies brings higher level of maturity in data processing. This frees up employee time, allowing them to focus on decision-making instead of analytics. Maturity in data management, in turn, directly affects the creation of value in the company.

In contrast to the highlighted advantages of implementing BI systems, there are a number of risks and limitations [11]:

- Need for high quality of initial data, for the correct operation of the system
- distrust of BI systems or lack of understanding of the principles of their operation, fear of unsuccessful implementation and, as a result, unwanted losses
- lack of clearly defined business goals for the implementation of the system and low involvement of managers.

3.2 Methodology for Analytical BI Reporting

3.2.1 Monitoring of the Main Indicators of the Oil Industry

To build BI reporting, analysis of the business customer's subject area was carried out. In order to effectively develop dashboards, a structured approach is required. An algorithm collects requirements for dashboards, based on the "Business Model Canvas" by Osterwalder and modified by Bunin, head of the visualization team at Yandex can also be used to develop BI reports.

The algorithm consists of the following steps [23]:

1. Determine the team
2. Conduct an interview with the customer to determine the task
3. Make a portrait of the user
4. Determine the format of user interaction with the dashboard
5. Describe the issues and business decisions that will be made

6. Identify data sources and data structure
7. Develop the main blocks of the report, select types of charts
8. Develop the layout and prototype of the dashboard
9. Carry out launch and testing.

For clarity, these steps can be visualized as a canvas. All elements are concisely described and are in one place, so it is most convenient to use the algorithm in this format [20].

Table 3.1 provides a pattern for gathering all the necessary information before starting to develop a dashboard.

Let us analyze each step in more detail and conduct research on a specific example [20]:

1. *Team*

This section outlines the roles that are involved in the dashboard creation the employees who perform these roles and their responsibilities. For effective interaction, a common online space for collaboration is created, chat and task management portal, with the containing and documentation).

Intended roles involved in the development of the dashboard include:

Customer, Project Manager/Product Manager, Data Engineer, BI Analyst/BI Developer, Designer and, Business Analyst.

2. *Understanding the task*

This section describes the activities of the business customer and outlines the basic business processes in the department under consideration.

The purpose of building an analytical dashboard is described and what needs it covers.

The business process and the tools currently used for the solution of this task are described.

3. *Main Users*

This specifies the key users who interact with the dashboard. At the same time, their roles and goals are spelled out.

4. *Format of user interactionwith the dashboard*

It describes the conditions under which users will interact with the dashboard and specifies technical conditions e.g., the dashboard must be displayed on large screens etc.

Table 3.1 Dashboard Canvas template for requirements collection

<table>
<tr><td>1. Team</td><td rowspan="2">7. Format and list of charts</td><td rowspan="3">2. Understanding the task</td><td>9. Testing and support</td><td rowspan="3">3. Users</td></tr>
<tr><td>6. Data</td><td rowspan="2">4. Context and format of interaction</td></tr>
<tr><td colspan="2" rowspan="2">8. Layout</td></tr>
<tr><td colspan="3">5. Business solutions and issues</td></tr>
</table>

5. *Questions answered by the dashboard and business decisions made*

A list of questions is created that users will receive answers to when interacting with the dashboard, and further actions are assumed. Based on the formed questions and actions, appropriate visualizations are selected.

6. *Datasources and datastructure*

Identify the data needed to calculate the metrics, where the data will come from, and how the data will be generated. The description of the business process can be in the User Story format.

In other words, it is important to find out the sources of the data and what data they can provide. Next, describe what the data structure will be, for example, a list of calculated metrics and their definitions.

7. *Select Basic Report Blocks and VisualizationTypes*

At this stage, based on the previous blocks, the visualization suitable for solving the problem is selected. If necessary, the report is divided into blocks and subblocks

8. *Report layout*

The report layout is designed in any suitable visual tool: Figma, Excel, depending on the level at which the layout is submitted for approval.

9. *Implementation of launch and testing*

Once the dashboard has been implemented, it is critical to study how people use it and how convenient it is for them to work with it. The simplest way to do this is to observe employees working with the dashboard or to create a survey to collect feedback from employees.

A feedback system collects comments on dashboards: for example, a chat with support or feedback form to report bugs.

If necessary, dashboard usage statistics are also monitored. Optionally, a list of Success Metrics for the dashboard can be generated.

3.2.2 Main Types of Dashboards and Data Visualizations

People often tend to perceive data more visually. In addition, visualization has the following advantages [19]:

- Focuses on key information
- Simplifies the analysis of large amounts of complex structured data
- Reduces the information load and focuses attention
- Ensures transparency of the data presented
- Contributes to the identification of relationships, patterns, errors in data
- Contributes to better perception due to better aesthetic appeal.

In order to conveniently present a large amount of complex data, analytical panels or dashboards are developed.

A dashboard is a combination of tables and charts that integrate and display heterogeneous analytical data. The advantages of building dashboards include:

- displaying the necessary data in one interface, which eliminates the need to switch between different windows
- Interactivity of visualizations: the ability to modify the visualization according to the needs of users
- Ability to compare and contrast data
- Additional insights through a combination of visualizations.

There are three types of dashboards [19]:

- Strategic: display indicators at a high level
- Analytical: display trends and dependencies in data in sections for several periods and allow for more flexible analysis
- Operating rooms: show detailed information, usually in real time, display deviations.

The right choice of data visualization is key to management decisions as it answers the analytical questions. There are a variety of different graphs that can be used to display certain data and help in conducting analysis, identifying problem areas, and finding growth points [19].

The following are the main types of visualizations used when building dashboards:

1. Table
2. Summary table
3. Line Graph/Time Line Graph
4. Bar chart
5. Combined schedule
6. Pie chart / "donation" diagram
7. Widget
8. "Sensor" diagram.

3.3 Development of Analytical Tool

3.3.1 The Profile of the Company

The company is an oil production enterprise specializing in oil and gas in the Russian and the CIS market for more than a decade.

Main services provided:

- Hydraulic fracturing
- Drilling wells
- Sidetracking

- Repair and cementing works
- Coiled tubing operations in wells.

The company is one of the leading providers of the above services.

The geography of activity includes such cities as Kogalym, Khanty-Mansiysk, Nizhnevartovsk, Noyabrsk, Nefteyugansk, and Samara, etc.

The company has its own laboratories, where the selection of materials used during the work is carried out.

In its production activities, the company uses modern approaches, advanced technologies and equipment, and constantly improves production processes and the qualifications of personnel.

The first primary source of data is the user interface of the corporate portal, where employees enter data on a daily basis. Users enter, modify, or delete data on the duration of events, the duration of downtime, the causes of downtime, the amount of proppant injected and fluid injected, the number of well operations performed, and more.

The second main data source is the 1C system, which provides data on personnel and equipment operated. Data from this system is consolidated into the appropriate PostgreSQL tables using a specially designed connector.

Algorithms written in SQL create materialized views in the database. Each materialized view contains an aggregated selection of data based on the existing tables in the database. An automatic data update mechanism written in Python is implemented for the materialized views. Each materialized view is periodically updated hourly.

The database is connected to the Apache Superset BI system, which builds dynamic reports based on materialized views from the PostgreSQL database.

The developed dashboards are integrated into the business customer's system and used by managers to make management decisions.

3.3.2 Business Processes: "as is" and "to be"

Fast access to data is needed to make management decisions. The faster the data is consolidated and aggregated, and the more accurate it is, the better project managers can analyze activities and potentially gain a competitive advantage in the marketplace.

First, let us look at the current business process for data collection in this company. The employees keep reports using Excel. At the beginning of each day, the administrator of the project office sends out an email notification about the need to provide data for the previous day for each of the 16 fleets. Over the next few hours, the administrator waits for the requested information to be provided, while at the same time starting to consolidate and process the data received. After the 15:00 deadline, the administrator prepares the draft of the Excel file containing the data and makes a preliminary mailing. Until the 17:00 deadline, the administrator awaits comments from the data providers in a reply letter and makes adjustments as they come in. At

18:00, the administrator sends out of the final Excel file containing the statistics for the previous day.

The advantages of this business process: the tasks performed do not require complex technical skills other than basic knowledge of MS Excel.

The disadvantages of this business process: the manual consolidation and processing of data from different Excel files requires the involvement of a specially dedicated person—the project administrator. The process is so time-consuming that it takes a whole working day, while the human errors cannot be excluded. In addition, the final report contains data relevant to the previous day, which prevents the timely provision of analytical information and quick response to the emergence of potential crisis situations.

The current business process is therefore sub-optimal and needs to be improved by introducing additional information technologies.

Let us consider the target business process after the implementation of the BI system. In the process of automating reporting, a transition is made from reporting using Excel tools to a digital application (corporate portal), where users enter data that is then displayed on interactive dashboards.

Each user now logs in to the corporate portal and then has two options: either go directly to viewing data on the dashboards or enter new data/change existing data using the data entry interface.

The data is refreshed at the beginning of each hour. Auto-update of data has been implemented to improve the speed of the dashboards.

When data is entered or changed, it is consolidated in the database, after which algorithms create the necessary slices, i.e., materialized views with auto-update often once an hour. Based on the materialized views created, BI reports are built, which are used by managers to monitor project activities.

The disadvantages of this business process:

- Implementing a BI system, requires technical experts to work with databases, develop BI reports, etc.
- After the development of the tool, it is necessary to train end users.

The advantages of this business process:

- Thanks to the creation of a single information and analytical space, it will significantly speed up the process of consolidating, processing, and providing data. This will eliminate the need for manual performance of these tasks from the business process chain and enable direct viewing and analyzing data
- Transparency and reliability
- Users are provided with a more convenient, easy-to-use analytical tool due to its interactivity and visibility
- Simultaneous round-the-clock access to the viewing tool
- Access from various types of devices: PCs, tablets, and smartphones.

The scheme of the target business process is shown in Fig. C.2 of Appendix C.

Based on the above requirements collecting algorithm, business needs analysis. As a result, the requirements for each dashboard were formed and the dashboards were developed (see Appendix C):

1. Results for all fleets
2. Results for individual fleets
3. Control of daily reporting
4. Lessons learned
5. Status of Fleets
6. NPE
7. KPI
8. Master's Desktop
9. Results for the day
10. Results for the month
11. Results for the year.

3.4 Conclusion

In this chapter, analytical reporting was developed using a Business Intelligence system to improve the quality of management of the oil production process.

The following tasks were performed:

1. Review of the main oil production methods, with a focus on hydraulic fracturing technology, which has proven to be an effective method for improving the efficiency of extracting hard-to-reach oil reserves
2. The business customer field of activity, identifying the main areas of activity, geographical presence, and key indicators used by managers to make decisions
3. Analysis of the market for technological solutions of the Business Intelligence class was carried out and the category of open-source solutions was chosen in order to modify the source code if necessary and eliminate additional costs for the purchase of the product and commercial licenses. Based on the results of the analysis, it was decided to implement the project using the Apache Superset system, since the characteristics of this tool met the key requirements
4. The methodology for collecting requirements for BI reports "Dashboard canvas" was chosen, since it offers a structured approach to analyzing the needs of a business customer. On the basis of this methodology, reporting requirements were collected and described
5. A set of BI reports based on the selected Apache Superset tool was developed, including eleven interactive dashboards
6. The project outcomes improved decision-making and operational efficiency, increased revenues, and reduced expenditure.

References

1. Anthopoulou, T.: Rural women in local agrofood production: between entrepreneurial initiatives and family strategies. A case study in Greece. J. Rural. Stud. **26**, 394–403 (2010)
2. Artificial Societies and Social Simulation using Ant Colony, Particle Swarm Optimization and Cultural Algorithms Source. In: Korosec, P (ed.) New achievements in evolutionary computation, p. 318. INTECH, Croatia. Downloaded from SCIYO.COM. ISBN 978–953–307–053–7 (Accessed 14.05.2025)
3. Bäck,T., Schwefel, H.P.: An overview of evolutionary algorithms for parameter optimization. Evol. Comp., **1**(1), 1–23 (1993)
4. Bäck, T.: Evolutionary algorithms in theory and practice. Oxford University Press, New York (1996)
5. Baluja, S., Caruana, R.: Removing the genetics from the standard genetic algorithm. In: Prieditis, A., Russell, S. (eds.) Proceedings of the 12th international conference on machine learning (ML-95), pp. 38–46. Morgan Kaufmann, Palo Alto, CA (1995)
6. Beightler, C.S., Phillips, D.T., Wilde, D.J.: Foundations of optimization, 2nd edn. Prentice Hall, Englewood Cliffs, NJ (1979)
7. Blum, C., Roli, A.: Metaheuristics in combinatorial optimization: Overview and conceptual comparison. ACM Comp. Surveys, **35**(3), 268–308 (2003). ISSN: 0360–0300. CODEN: CMSVAN http://iridia.ulb.ac.be/~meta/newsite/downloads/ACSUR-blum-roli.pdf (Accessed 14.05.2025)
8. DeJong, K.A.: Genetic algorithms are NOT function optimisers. In: Whitley, L.D. (ed.) Foundations of genetic algorithms 2, Morgan Kaufinann, (1993)
9. Dorigo, M., Stützle, T.: Ant colony optimization, a bradford book. The MIT Press Cambridge, London, England, MA. ISBN 0–262–04219–3
10. Dorigo, M., Gambardella, L.M.: Ant colony system: A cooperative learning approach to the traveling salesman problem. Technical Report TR/IRIDIA/1996–5, IRIDIA, Université Libre de Bruxelles, (1996)
11. Eiben, A.E. (2002). Evolutionary computing: the most powerful problem solver in the universe? Dutch Mathematical Archive (Nederlands Archief voor Wiskunde), 5/3(2), 126–131, 2002 http://www.cs.vu.nl/~gusz/papers/ecintro-naw.ps (Accessed 14.05.2025)
12. Fogel, D.B.: Evolutionary computation. IEEE Press, (1995)
13. Fogel, L.J., Owens, A.J., Walsh, M.J.: Artificial intelligence through simulated evolution. Wiley, New York (1966)
14. Streichert, F.: Introduction to evolutionary algorithms, presented at the Frankfurt Math Finance Workshop, (2002)
15. Fletschner, D., Carter, M.R.: Constructing and reconstructing gender: reference group effects and women's demand for entrepreneurial capital. J Socio- Econ **37**, 672–693 (2008)
16. García, M.-C.D., Welter, F.: Gender identities and practices: interpreting women entrepreneurs' narratives. Int. Small Bus. J. **31**(4), 384–404 (2013)
17. Goldberg, D.E.: Genetic algorithms in search optimization and machine learning. Addison-Wesley, Reading, MA (1989)
18. Goltz, S., Buche, M.W., Pathak, S. (2015). Political empowerment, rule of law, and women's entry into entrepreneurship. J. Small Bus. Manag. **53**(3), 605–626 (2015)
19. Glover, F., Kochenberger, G.A. (eds.): Handbook of metaheuristics, volume 57 of International Series in Operations Research & Management Science. Kluwer Academic Publishers/Springer, New York, USA (2003)
20. Hillier, M.S., Hillier, F.S.: Conventional optimization techniques. Chapter 1. In: Sarker, R., Mohammadian, M., Yao, X (eds.) Evolutionary optimization, pp.3–25. Kluwer Academic Publishers, (2002)
21. Yao, X.: Evolutionary computation: A gentle introduction. Chapter 2. In: Sarker, R., Mohammadian, M., Yao, X. (eds.) Evolutionary optimization, pp.27–53. Kluwer Academic Publishers, (2022)

22. Yarskaya A.A.: Essence and role of BI-systems in the modern economy. (2021)
23. Bunin, R.: How to create useful dashboards for business: algorithm, principles of layout, tools, architecture https://sense23.com/post/kak-sozdavat-poleznye-dlya-biznesa-dashbordy-algoritm-printsipy-verstki-instrumenty-arhitektura?ysclid=lw61qhza58678759521 (Accessed 14.05.2025)

Chapter 4
Harnessing Diversity: Intelligent Portfolio Management for Financial Group

Abstract This chapter looks at digitalization in large financial environments including such powerful multinationals such as Raiffeisen. Their organizational complexity, a potential crisis trigger, is compounded by extremely fierce competition and some other negative factors. Consequently, portfolio management for such complex environments should be based on a thorough multi-criteria optimization. Therefore, the supporting original techniques and tools integrate intelligent large-language modeling, sentiment analysis and visualization engine. The suggested digitalization pattern, implemented as an API, significantly improves the prediction accuracy, while the system remains scalable and robust. Customer satisfaction was confirmed by a post-production survey.

4.1 Introduction

Stock market financial instruments have historically formed an important part of mankind monetary relations. Depending on the historical period, the popularity of some of them has increased: for instance, consider the situation in Russia in the 1990s, when, due to extreme devaluation, the threat of default and almost zero credibility of the Central Bank and the national currency, the biggest players on the Russian market used treasury bills of the most important companies to settle among themselves: the oil giant Yukos alone issued treasury bills worth tens of billions of conventional units.

Today, financial institutions in Russia are actively developing, confidence in the Central Bank and the national currency is at a much higher level than in the 1990s. The government stimulation of the stock market development has led to the creation of a large number of investment services and support for this process. According to the official data of the Moscow Exchange, the number of registered investors exceeded 25 million people in 2023, which means that every sixth person in Russia has a brokerage account and thus access to trading in financial instruments.

The graph shows the growth of registered investors on the Moscow Exchange from 2018 to 2023:

S. V. Zykov, *IT Crisisology Patterns and Practices*, Smart Innovation, Systems and Technologies 449, https://doi.org/10.1007/978-981-95-1260-7_4

Consequently, there has been a significant inflow of individual investors to the Exchange over this five-year period. This happened as a result of the development of investment services, products, and conditions such as:

- Significant regulatory relief in opening and maintaining a brokerage account. The government introduced regulations requiring brokerage firms to become tax agents for their clients. This made it much easier for the individual investor to report to the tax authorities. The procedures for opening a brokerage account, registration, and authorization of a user on the stock exchange have been greatly simplified. As a result, a person wishing to open a brokerage account can do so in 15 min with only a passport.
- In 2015, the government approved a bill on the introduction of individual investment accounts, which allowed not to pay tax on income from portfolios consisting exclusively of securities of Russian issuers.
- Brokerage companies based on structured banks began to appear. Tinkov Investments was a pioneer in this field and is still the largest bank-based brokerage company. Let us quote Oleg Tinkov, the founder of Tinkoff Bank: *"As a clientof the bank, I want to buy Apple shares from my Apple smartphone"*. The integration of brokerage services with banks has made it possible to virtually eliminate the user authorization process and give access to the mobile infrastructure, i.e., the use of brokerage services in mobile app.

All this could not but stimulate a huge inflow of individual investors to the Moscow Exchange. The promotion of financial literacy, capital management courses, and the accumulation of experience of living in a credit and market economy also played a significant role. Citizens are increasingly reluctant to rely on state support after their working lives have come to an end and are increasingly taking care of their own pensions.

4.1.1 Overview of the Russian Market and a Mutual Funds

The most common financial instrument for trading among individual investors is the shares of Russian companies. According to the Moscow Exchange, more than 75% of all trading operations are carried out exclusively on these instruments. This can be explained by the ease of understanding how the instrument works, its liquidity, availability, flexibility in portfolio composition, diversity and marketing component.

However, the consequence of such a huge influx of new players into the market is a significant decrease in the average professional qualification of an investor. Attending specialized courses does not make a person a professional trader or a professional portfolio manager. Such skills are acquired through experiencing a large number of market downturns and upswings. This is why beginners lose enthusiasm after their first "loss" and stop investing altogether. Others decide to entrust the management of their capital on the stock market to a specialized company, where a team of experienced professionals decides on the management of a portfolio of financial instruments.

Even before the "investment boom" in Russia, investment management organizations began to develop to provide capital management services to wealthy clients. Soon, however, there was a demand for such services from people with more modest means. As a result, investment companies began to develop products such as mutual funds. A mutual fund is an investment product offered by a management company (MC), which consists of building a portfolio of securities and issuing units that represent the right to own a portion of the capital, which is operated by the MC with the aim of increasing it.

Today, mutual funds are a safer alternative to self-investment because:

- Asset management is carried out by a team of professional investment managers and is based on the analysis of the entire MC, which is inaccessible tom small Investors.
- Each fund has its own investment profile: target yield, target risk level, preferred economic sectors of the issuing companies, preferred securities, etc. A wide range of products is available to satisfy investors with different needs.
- The portfolio management fee may vary from one investment product to another within the same MC with an average of 3–4%.

Today, the Russian investment products market is at a fairly developed level. Most of the players have been in the market for more than 15 years, which indicates a well-developed industry infrastructure and established competition.

The Russian MC market is characterized by monopolistic competition, which is reflected in such companies as Raiffeisen Capital, Alfa Capital, Aton, Promsvyaz Investments, VIM Investments, and VTB Capital. These companies account for more than 85% of the market for open investment products. The remaining 15% is made up of small, specialized, niche players that traditionally manage the capital of a limited number of people or one client at all. These are almost always closed products with no access for the average individual investor.

The most popular investment products for the monopoly part of the market are mutual funds. The range of mutual funds offered by Russian MCs is wide enough to satisfy any investor's needs. Traditionally, the most popular portfolios are made up of financial instruments, such as stocks and bonds, which are familiar to all investors, even without special professional qualifications. Since 2015, thanks to government support, portfolios made up exclusively of Russian issuers have taken the lead.

Based on this parameter, let us consider the portfolios of the largest Russian MCs, which consist exclusively of shares of Russian companies. The criteria for selecting mutual funds include:

1. The fund portfolios must predominantly consist of shares of Russian companies. 'Predominantly' means more than 90%, while the presence of other securities, such as bonds and valuable exchange-traded assets in the form of precious metals, and free cash is allowed. The presence of other financial instruments is explained

by the need for immediate liquidity of the instruments, the fulfillment of obligations without extraordinary rebalancing of the portfolio, and the minimal need for hedging assets.

2. The amount of the fund's net assets. This is the main indicator of the size of the fund, the more capital, the more investors, and therefore the more popular the fund. Of course, it has certain limitations:

   2.1. Legal restrictions, QS according to Russian, mutual funds are prohibited to have a majority stake in the company and participate in its management. Unfortunately, in order to comply with this restriction, big mutual funds are forced to diversify their portfolios as much as possible, which results in lost profits from not trading a "good" asset.
   2.2. Low liquidity at critical times. Despite the fact that they only legally own a minority stake in the company, the securities packages can be quite large, so that there is an illiquidity crisis in the event of an emergency closing of a position. An example of this is the panic on the Russian stock market on February 24, 2022, when all players began to close their positions on an emergency basis. On the afternoon of February 25, the Moscow Stock Exchange suspended trading for more than three weeks. Most funds had no time to react as they simply could not find buyers for such a volume of securities.

In addition, the overwhelming majority of users of investment products prefer mutual funds from the top 10 NAV rating. It is most relevant to consider these portfolios, although they are far from ideal.

Availability of reporting. In the RF, management companies are obliged to provide monthly reports on their products in a uniform reporting form. However, not all MCs provide reports in a convenient format.

Based on these criteria, the following unit investment funds were selected for consideration:

1. "Liquid Shares" from MC Alfa Capital
2. "Shares" from MC Raiffeisen Capital
3. "Pyotr Stolypin" from MC ATON
4. "Perspective Investments" from MC Promsvyaz.

It should be noted that the selection was made from portfolios consisting mainly of shares of Russian companies. There are funds on the market with portfolios of larger capitalization, but they do not meet the criteria set and cannot be used in the work.

The best indicator of portfolio efficiency is always the yield relative to the market average. Let us investigate the behavior of fund portfolios over three time periods:

1. 2021, economic recovery from the coronavirus pandemic, presence of relative calm and cautious optimism in the market
2. 2022, the most critical and intense year for the Russian stock market. For more than three weeks, the Moscow stock exchange was not trading, while the market fluctuated from a catastrophic decline to a rapid recovery

3. Early 2023, acceptance of the new reality, moderate market recovery.

Analyzing the behavior of portfolio returns over these intervals not only unambiguously identifies the most effective portfolio management, but also highlights a relevant candidate for further research (see Fig. D.1 of Appendix D).

The chart shows the previously selected portfolios and their comparison with the average market yield. It is possible to identify the favorite in this time picture, namely UIF Peter Stolypin from MC ATON represents the most attractive profitability which, is considerably higher than average market profitability on all analyzed intervals.

MC ATON often declares that it uses the latest techniques for market analysis, combining classical fundamental analysis with computer algorithms, for the analysis of technical indicators. The company spends considerable funds on staff training and the improvement of analysis methods, which is ultimately billed to the client in the form of the highest commission for MC ATON. The strategy of MC ATON is rather complex, as standard practices of portfolio management are inseparably mixed with the newest technologies.

Other companies are more conservative in their campaigns, and they, unlike MC ATON, declare that portfolios are managed by professional managers, and all decisions are made on the basis of their experience.

The behavior of the portfolios of Raiffeisen Capital, Promsvyaz, and Alfa Capital is very similar, so much so that their profitability graphs are almost identical and hard to distinguish. All of them failed in their main task of exceeding the average market yield in the interval from 2021 to 2023, and in February 2022 the graph of yield became less than the average market yield and could not reach the level of 2021.

The return graphs of all funds have significant similarity with the market average. This is a consequence, of an excessively high diversification of portfolios, which include most of the companies listed in the index of the Moscow Exchange. This input does not allow to noticeably increase the portfolio yield in relation to the market average. The portfolio of each fund contains securities, on average, from 25 to 40 issuers. Such diversification in the Russian market also significantly complicates the process of analyzing and making decisions on portfolio management. Therefore, in order to compete in the capital management market, it has become vital to use alternative sources of analysis using machine learning algorithms.

4.1.2 Portfolio of Raiffeisen Capital Management Company

The mutual fund "Shares" of Raiffeisen has a long history and world famous. This fund is also interesting because Raiffeisen, in its advertising policy, appeals to the fact that it is an organization with a long history and world fame. The name refers to a famous Austrian brand, which creates an impression of using the best world practices in the company's work.

UIF "Shares" consists exclusively of shares of Russian issuers and issuers that are de facto domestic companies of the Russian Federation.

The Equity UIF is Raiffeisen Capital's flagship product and therefore the most popular. First of all, let us consider the sector diversification of the portfolio. It is important to note the presence of an unequal share of any industry or its absence (see Fig. D.2 of Appendix D).

As one would expect, the most popular sector in the portfolio is oil and gas, which occupies, on average, slightly less than half of the portfolio. This does not mean that the portfolio is poorly diversified, because about 40% of the Russian economy is occupied by the production and export of hydrocarbon products, and together with post-products, this share increases to 60%. Consequently, the companies in the Russian Federation with the largest capitalization and opportunities to acquire a minority stake are the players from this industry. Unfortunately, to date, companies from this industry are market outsiders in terms of profitability and expectations. A build-up of stakes in the oil and gas industry during the uncertainty of the first half of 2022, at which time companies from this industry were seen as a defensive asset that would be the last to fall under foreign restrictions. However, hopes were not realized and management started to move assets to more buoyant areas of the economy.

We see the maintenance of fairly high positions in the following sectors: consumer staples—17% on average, financials—15%, energy—13%.

Consumer staples are now seen as a protective asset due to their independent operation in the Russian domestic market, the existence of almost perfect competition in many categories of goods, and stable constant and predictable demand for the products. Securities of such companies, although not showing amazing growth, are a safe haven relative to companies from other industries.

The finance sector is the driver of stock market growth. Securities of SBER, VTB and TCS have been leaders in terms of trading volume on the Moscow Exchange for a long time. They provide unprecedented yields as a result of changes in the value of securities, as well as dividend payments. SBER's growth of more than 300% since the beginning of 2022 is a precedent for the Russian stock market as a blue-chip company that has shown such returns.

Also of interest is the presence in the portfolio of companies from the precious metals and stones industry. It is well known that these commodities are historically a classic hedging tool, i.e., asset protection, because in cases of market panic, crisis and other unpleasant situations for the stock market, the price of such commodities goes in the opposite direction, and therefore, the securities of companies from this industry follow the price of precious metals.

Note a sharp increase in the share of securities of companies from the transport sector in late 2023–early 2024. After the beginning of the period of uncertainty, the assets of the transport industry, the flagship of which are air carriers, became outsiders, due to the imposed sanctions, limitation of the flight zone, ban on maintenance of aircrafts and the purchase of new ones. However, thanks to the state financing of aircraft building projects, the domestic aircraft building industry began to demonstrate fairly good results. The sharp increase in the share of presence in the securities of the industry can be explained by the long-term confidence in the prospects of airline companies.

Other sectors, such as the chemical industry, telecom and IT, are not of special interest for the fund and, in total, occupy less than 10% of the portfolio.

According to the sectoral diversification of the portfolio, despite a presence in all sectors of the Russian economy, the portfolio is balanced in terms of diversification ratio and focus. 90% of the portfolio is occupied by a presence in 5 industries. No industry occupies more than half of the portfolio. The dominance of the oil and gas sector is explained by the specifics of the Russian economy. In general, the portfolio can be assessed as restrained and focused within the limits of the portfolio's opportunities with high capitalization.

4.1.3 Portfolio Management and Rebalancing

The research basis included publicly available information that management companies are obliged to disclose. The document "Statement on the amount of net assets of the fund" is published monthly on the official website of the MC. The structure of this document contains information about where the capital is invested in the securities of Russian issuers.

Having these documents let us build the capital structure contained in the structure of the fund's assets and evaluate its change.

Based on the previous analysis of the fund's industry composition, the industry structure has not undergone a major change that requires attention. The sectoral structure roughly repeats the Moscow Exchange index, and it makes no sense to assess the sectoral rebalancing.

Evaluating rebalancing by companies within the industry composition is more applicable, illustrative and suitable for its comparison with the fund's performance indicators. These indicator will allow to assess how portfolio management activity has changed over different time periods. As mentioned earlier, the time horizon of the review represents two unique periods in the Russian economy: a period of uncertainty and a period of new reality. The rebalancing indicator will allow to see how the management of the management company acted in different time intervals, especially in areas of particular turbulence.

This study will make it possible to assess the portfolio behavior of the MC, identify and describe the fund management strategy. It will also allow making assumptions about the use or non-use of technology in the decision-making process. This can be assessed on the basis of:

1. The fund's repeatability to the index. There are many funds that set themselves the task to correspond to this or that index, but they directly mention it in their marketing campaigns. As a rule, such funds have a minimal commission, a huge number of investors and, in case such mutual funds are exchange-traded, significant opportunities for unit liquidity. However, large funds, whose goal is not to match the market index, but to outperform it, often use this structure as a base,

modifying it and adapting it to market situations. If the fund's structure overcomes a reasonable level and practically repeats the index, it indicates inaction of the management and "insurance" of profitability to the depositors. In such a case, the fund managers do not deservedly take a higher commission and do not fulfill their obligations. The repetition of the fund's asset structure to the market index also indicates the unwillingness or inability of the management to use analyzing technologies for decision-making. If the fund is unwilling to invest in its own team of investment analysts, to maintain a quality personnel policy, team building, professional development, best practices, etc., the result, in the form of final investment analysis, is of low quality and causes distrust of portfolio managers. This situation leaves no choice but to rebalance the market index.

2. The rebalancing percentage is too small. It is known that long-term investing and short-term trading are completely different approaches with different goals. Modern technology allows optimizing the decision-making process, analyzing and obtaining information and taking into account more factors. An effective fund should include, elements of trading, i.e., fast, short-term trading in order to get much higher profits in a short period of time. This does not mean that all funds should turn into speculative organizations, where specialists in trading and technical analysis "throw dice" several hundred times a day. However, missing the opportunities provided by technology, in terms of trading and speculation, leads to market inefficiency of the fund. Today, even the most conservative funds divide their portfolios into a retained portion (usually more than 75% of the portfolio, and is made using classical, "manual" fundamental analysis), and a speculative portion (usually no more than 25% of the portfolio, the capital is managed by a more qualified team of analysts, whose work is based on the use of computer algorithms and rapid buying and selling). This meets the demands of investors in terms of asset protection, but also allows, based on an acceptable level of risk, to obtain a return significantly higher than the market average. In this regard, a negligible rebalancing percentage may indicate that this approach is not being utilized.

To evaluate the performance of the fund, let us use the rate of return. It is possible to use more informative and useful indicators, such as Sharpe ratio and its analogs: Alpha Jensen ratio, Modigliani ratio, Traynor ratio and others. However, their calculation requires an understanding of the market risk indicator for each asset in the portfolio. We believe that the classical measure of risk as the average standard deviation of an asset is not relevant in the current environment. After the beginning of the volatility period, assets that were characterized by extremely low volatility became one of the riskiest assets on the market, due to this fact, the classical risk measure should be further developed and will not be used further.

Profitability is the only indicator of investment efficiency that is interesting for an average investor, therefore the most relevant assessment will be made only on the basis of this indicator.

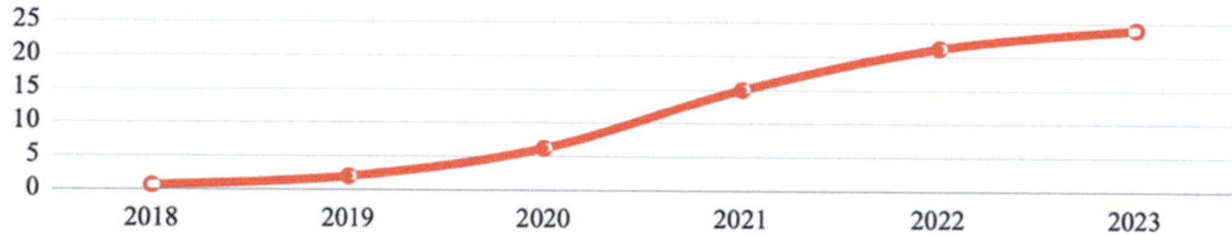

Fig. 4.1 Number of registered Moscow Exchange investors (2017–2023), mln

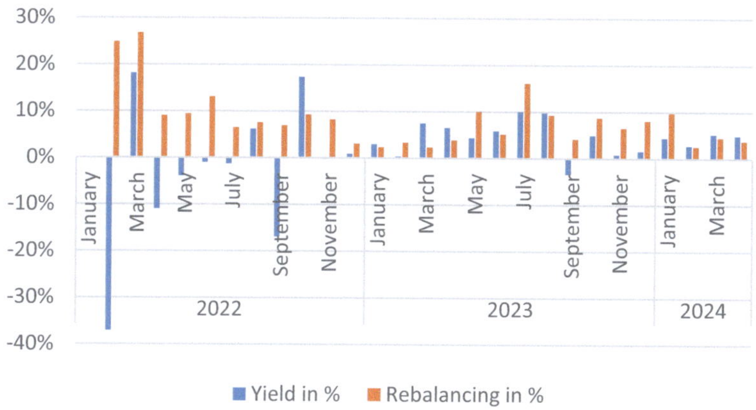

Fig. 4.2 Rebalancing and yield, %

Linking the amount of rebalancing and profitability will allow to draw conclusions regarding portfolio management from the diagram of the relationship between rebalancing and profitability (see Fig. 4.2).

Since the indicators are measured in real terms, rebalancing and profitability for January is taken as zero.

Analyzing the chart at a glance, there is fluctuations from a sharp increase in the indicators to keeping them almost at zero. At the beginning of the chart, in February 2022, at the very beginning of the crisis period, there is a huge % rebalancing and proportional losses. On February 24, after the announcement of the SMO, stock players had no more than 24 h before the stock market shut down. Some insiders had this information, while others guessed about the imminent closure of trading and took action following the excitement of other market players. The desire to move their assets to calmer areas, which will not be so affected by hostilities and sanctions restrictions, led to a large % of portfolio rebalancing. This % could have been much higher, but the hasty closing of trading on the Moscow Exchange cut off all limit orders, and sell orders could not be executed. However, regardless of the sphere, the whole stock market of the Russian Federation, under the pressure of the panic of market players, catastrophically flew downward. This situation was frozen for more than three weeks, until March 18, when trading on the Moscow stock exchange resumed. Economic and geopolitical uncertainties in March were even a little higher than in February, but the theory of market fluctuations in this situation showed practical proof and the market, even against the background of the absence of positive news, began to correct toward higher yields. Also in March, the fund managers completed the "reconfiguration" of the portfolio. These two months show a large percentage of both rebalancing and yield index. However, these changes are purely due to market panic are irrelevant to this study.

From mid-Spring through to November 2022 being, we see increased portfolio rebalancing at 7–10% per month, but returns are negative during this period. This period is characterized by ongoing uncertainty and the related local oil and gas crisis. Thus, the European Union imposed sanctions limiting tanker transportation of oil and liquefied gas. World markets were feverish and the only possible strategy was to follow the market index, not to get a loss below the market average. Rebalancing of the fund during this period is characterized by bringing the structure of the fund in line with the market index. This did not happen overnight, but more smoothly during half a year. These dynamics can be traced both by sector analysis and analysis of specific issuer companies in the portfolio. Firstly, the 2nd and 3rd tier companies were excluded from the portfolio, the portfolio structure was reduced to the presence of only blue-chip issuers.

The year 2023 saw relative market calm. In the period from the beginning of 2023 to the present, no events have occurred in the market that would lead to panic and\or significant excitement in the market. Therefore, let us assess the impact of rebalancing on the fund's returns. Analyzing this period, the level of changes for the period varies from 4 to 15%. These are the peak values, in other intervals, the level remains at an average of 8–9%. Fund returns were relatively strong in the first three quarters of 2023, peaking at 10% in July. However, starting in the fourth quarter, returns fell sharply remaining still positive. The average return since the fourth quarter of 2023 has been around 3% monthly. The market average was 2%, meaning the fund met its entire primary objective and outperformed the market.

There is no evident correlation between the fund returns and rebalancing. The fund strategy in the period under review can be fully described as regular rebalancing, and as the shares of each sector will change insignificantly, the fund adheres to the same sector structure almost throughout the entire period of the crisis.

Following this strategy implies that the management adopts target shares of the sector structure of the portfolio and periodically revises it.

In the context of this strategy, computer analysis technologies, algorithms, and predictive models based on artificial intelligence are considered as auxiliary and are used to support the analysis based on the managers experience. Today, fund investors expect returns significantly higher than the market average. They are not ready to receive the market yield and pay the commission of the management company.

As mentioned earlier, MC today vitally need the introduction of technologies in terms of analyzing big data and building predictive models.

In order to satisfy the interests of the largest number of investors, it is recommended to divide the portfolio into a holding part (managed manually through fundamental analysis) and a speculative part (managed through predictive models and involving quick trading operations). Such a portfolio structure will maintain an acceptable level of risk of the portfolio as a whole, realizes the goals in terms of long-term investing, but at the same time will give the opportunity to gain profit due to short-term “imperfections” of the market.

Let us consider the development of a predictive model for analyzing news reports based on large generative models.

4.2 Technologies for a Predictive LLM Model

4.2.1 LLM Models and Agents

Large language models LM accept natural language input from a human and provide natural language output. All these models are based on ML and DL; thus, they can be used to perform and deliver commendable results for numerous tasks.

As highlighted previously section, the LLM evolution is connected with advancement and the emergence of new neural networks. Some of the first efforts are based on the Recurrent Neural Networks (RNN) and a particular type of RNN called Long Short-Term Memories (LSTM). These models were quite effective in managing text data but, as we have seen, there were some issues with the management of sequences and parallelism. The turning point was the Transformer Architecture outlined in the paper titled *Attention is All You Need* [6]. This work introduced the attention mechanism which made the models capable of processing long sequences of text, more than one sequence at a time.

Currently, LLMs can be divided into two main categories; open models and closed models. These categories have different features concerning their availability and usability. Let us give examples for each category.

1. BERT (Bidirectional Encoder Representations from Transformers): This model was proposed by Google and was one of the earliest models that tried to capture the bi-directional connection between text where the flow of information is two way. BERT has been designed as a generic model for text understanding and has been pre-trained on a very large text corpus that includes Wikipedia and BookCorpus. It employs a bi-directional language model which reads the text from left to right and also from right to left. It is easy to access and has the option of learning more, where by it is commonly used by academics and other firms.
2. GPT-2 (Generative Pre-trained Transformer 2): This model is by Open AI and was developed before the GPT-3,which is the latest one. The most significant aspect of GPT-2 is that it can generate text that is grammatically and meaningfully coherent in response to the provided input. It is in the public domain and this means that anyone can, at any time, decide to make changes to it.
3. RoBERTa (Robustly optimized BERT approach): Facebook AI has unveiled what is likely the next big advancement in BERT. RoBERTa has a few additional training and optimization options that allow better performance on a diverse set of NLP tasks.

4. GPT-3 and GPT-4 (Generative Pre-trained Transformer 3 and 4): developed by the OpenAI Company. The biggest and most efficient models with hundreds of billions of parameters for text generation and language comprehension. These models feature APIs that makes them handy; however, their application is limited.
5. T5 (Text-to-Text Transfer Transformer): An approach by Google that uses a very large corpus of text data and is able to solve any problem that can be stated as a text-to-text + problem. T5 can be obtained from Google Cloud and is therefore business applicable settings.

In order to develop agents using LLMs, there are several technologies that help the models understand data, learn, and function in different situations. A possible way of addressing this is the Retrieval-Augmented Generation (RAG—Retrieval-Augmented Generation) approach. This has the features of generative models to extract information from external sources. RAG assists the agent in understanding what data the agent requires to get in context; the agent can use the data to respond to the user query. For instance, if a user has a query related to recent events, the agent will be able to retrieve the latest information from news sources and provide the most relevant answer. This is useful for the agents to be more flexible and informed.

The LLM-based agents employ ML and DL techniques to analyze the data and present conclusions. The agents learn from the database provided and use this knowledge to make predictions, as well as decisions, through ML. DL is a form of ML that uses neural networks to learn from unstructured data, such as textual data or pictures. ML and DL enable the agents to learn from experience and from new data, which helps the agents to enhance performance. For instance, a form of ML known as reinforcement learning to learn how people likely respond to its actions, and then alter these actions in kind. Thus, scenarios and user demands can be dynamic, and agents' responses and actions can be refined and enhanced.

Agent is the main technology used in NLP. NLP enables models to comprehend what a human is saying and this is vital when it comes to the interaction between a model and a user. These include BERT and GPT, the modern NLP techniques used in converting text data the model to understand the context. NLP has some tasks including tokenization, parsing and semantic analysis, named entity recognition, and coreference resolution. These tasks explain the structure of the texts; thus, they give appropriate and meaningful responses. For instance, in text analysis, a model can choose which portions of the text should be treated as proper names, which words are actions, and which ones are objects, which in turn would enhance the model comprehension of the user query.

The construction of agents involves ML algorithms and analytical methods and techniques. These enable the agents to understand and analyze large data structures, relationships and occurrences to derive the data to make predictions and decisions. For example, in clustering and classification, the agents can group the information and act in a certain way, depending on the group. Other methods are the use of statistical models and optimization methods where the agents can determine the best possible solution. These techniques enable the agents to modify behavior even learn in new conditions.

Model tuning and pre-training is an important part building LLM-based agents. These processes allow to tailor models to specific tasks and improve their performance.

In the process of fine-tuning a model, there are several approaches including Supervised Fine-Tuning (SFT). SFT is a strategy that involves preparation of a dataset to train a model where the data instances have been manually classified. For instance, if the agent is to be employed in a customer relations service, it can be fine-tuned on real-life customer data. This process entails updating the weights of a model when it produces wrong answers to enable the model to learn and come up with better solutions. It assists in increasing the likelihood of the answers provided by the model being relevant and useful.

The second most effective approach is called Reinforcement Learning from Human Feedback (RLHF). This approach enhances performance according to the feedback by the user. The responses generated by the model are evaluated by the users and the resultant scores are then used to adjust the parameters of the model. In the case of RLHF, the model is able to learn from the feedback. This method is appropriate for the situations where users' preferences change frequently or in other words, dynamic environments.

All the above-discussed models of acquisition involve Transfer Learning approaches. This technique helps in transferring the learning from one model to another. For instance, a generative model that has been conditioned on a large corpus of text data can be fine-tuned with other corpus for specific goals. Transfer Learning saves time and costs training models and\or other adapting.

Another approach to pre-training is through the use of self-supervision. In this case, the model is trained on a dataset in which there is no clear distinction between the training set and the test set, and the model takes advantage of the inherent ordering of the data. This makes it possible to train models on the common data sets that are not tagged.

However, it is also crucial to consider the validation and testing of the tuning and pre-training models. To test the learning ability of the model and its applicability to a different set of data, models should be tested on different datasets. This process involves confirming the models and their assessment criteria such as precision, recall and the F1 measure.

These technologies and methods of tuning and pre-training form the basis for the efficient and versatile agents. When it comes to the application of RAG, ML, DL, NLP, and pre-training techniques, the agents are created to analyze the data, make decisions, and provide information to the users in real time. This makes agents effective in business process automation, recommendation systems, and customer support.

4.2.2 Practices for LLM-Based Predictive Models

Sberbank, one of the largest banks in Russia, started to apply LLMs to enhance the productivity and provide better services.

The key idea of LLM at Sberbank is to use ML to identify tendencies and, with the use of the obtained data, attempt to predict changes in the markets. It is developed on the basis of data of financial operations, announcements, statements and reports of companies, and macro rates. This process involves several key steps: The processes data collection, cleaning, modeling, and the incorporation of the model into the company activities.

The first stage is data collection. To prepare the macroeconomic forecast, Sberbank uses its databases, mass media, financial statements, and the platforms with the macroeconomic data. This data contains the transaction, the cash flow, the stock exchange variations, the news, and the reports. The reason for this is that the data in the model must be accurate and consistent.

Data cleaning is the process of removing the undesirable facts, and this entails the handling of missing values. This process is very time consuming and requires the use of computers and certain algorithms to work with big data. This data is masked in compliance with the privacy standards.

DL and reinforcement learning is used by Sberbank to DL and RL identify relations and heterogeneity of data.

After the training process, a model is assessed to check its efficiency in evaluation. This entails using the model in other datasets. Assessment was based on such parameters as correctness, inclusiveness and F1 measure. These metrics assist in identifying the model effectiveness. Another aspect is the interpretation of the results to identify which factors affect the model decision-making process, and how the decisions made by the model can be employed in operational business environments.

At this stage, it is necessary to integrate the model into business processes. At Sberbank the model is applied for predicting the fluctuations of prices in financial markets, the identification of risks, and investment strategies. The model is connected with the bank systems, which enable automatic data updating and real-time forecasting. This helps the bank traders and analysts to formulate better decisions and enhance their approaches. For instance, the model can forecast that a specific firm share price will rise in the next few months; subsequently, the bank will consider purchasing the stock.

It was found that the Sberbank LLM improves risk management. The model uses data on loan applications, customer financial status, and the current market situation, to establish the likely default rates. This enables the bank to minimize risk and decrease the total number of loans issued. Sberbank report indicates that LLM has reduced defaults by 20% and enhanced performance.

Also, the LLM booed organizational effectiveness and client satisfaction. The model performs the functions of customer requests and feedback analysis. It deals with textual data coming from communication channels, such as e-mail and chats, to develop a better understanding of bank clients and their needs to provide them with the most suitable solutions. This in turn enhances the customer experience and satisfaction. The LLM reduced query processing by 30% and raised customer satisfaction by 25%.

Another critical area of the LLM is combating financial crime. The model processes transactions and flags any activities which might be a sign of fraud or money laundering. This makes it easier for the bank to spot any potential threats and respond times. Due to the implementation of LLM, the reported cases of fraud incidence have been reduced by 18%.

Not only the companies, but also the industries benefit from LLM-based predictive models. McKinsey's research reveals that organizations that adopt LLM achieve an average improvement in their financial results of 5–10%. This is because of more accurate forecasts more effective risk and operations management. The companies better understand the market and provide the better services.

The results indicate the efficiency of LLM-based predictive models.

This chart illustrates the enhancements in different aspects of the Sberbank business, such as the forecast accuracy, portfolio profitability, default rate, response time to requests, customers satisfaction and fraud (see Fig. 4.3).

Sberbank's positive experience with the application of LLMs in financial analysis and risk management demonstrates the possibility of the enhancement of the efficiency and accuracy of these technologies in other fields. Other industries, for example medicine, are also utilizing this technology to improve the diagnosis and treatment of diseases. A specific example of this domain is the Cifromed company, which uses LLM to analyze medical data and forecast the likelihood of getting chronic diseases.

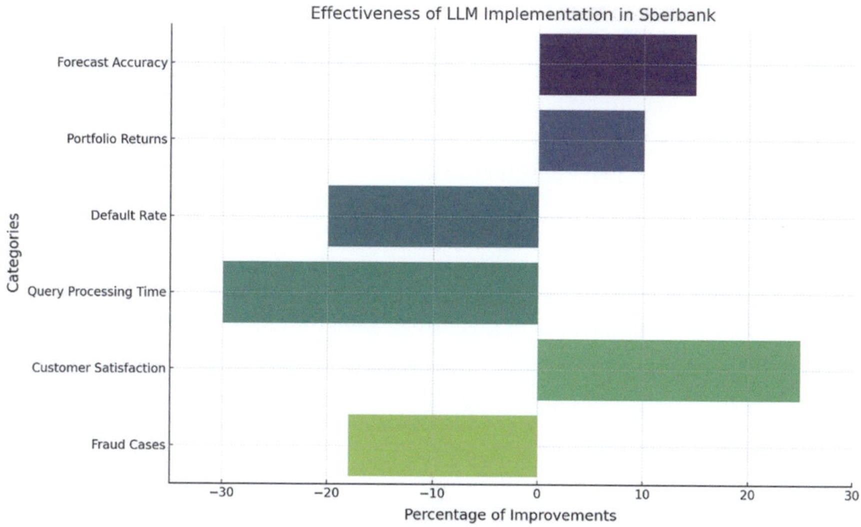

Fig. 4.3 Effectiveness of LLM implementation in Sberbank

The primary goals of the LLM at Cifromed are preprocessing of massive medical data, pattern recognition, and diseases prognosis. Based on this, the model is fed with medical history, test results, genetic data, and information about patient lifestyle. This process involves several key steps: data collection and preparation, model building, assessment, and application.

The initial and critical part is the data collection process. The data for Cifromed is acquired from various sources, such as electronic health records, laboratory results, prescription drugs, and DNA. This data contains a patient's diagnosis, test results, genetic mutations, and other vital data that determine the patient's state. This means that the data being used should be consistent, complete accurate and comprehensive because this will help to determine the credibility of the model's predictions.

It is the process of preparing data by cleaning and removing the noise, outliers, missing, and inconsistent items and converting them for training the model. This process is very computationally intensive and, therefore, contains techniques to analyze big data. Another key point is data depersonalization to protect confidentiality.

Model training starts after data preparation. Cifromed applies deep learning along with RLHF to develop models that can recognize the data dependencies and patterns.

Deep learning forms multi-layer neural networks which can process text and comprehend context. RLHF enables the model to learn from the changing conditions and fine-tune its predictions with the help of physicians' feedback. It is based on patient data; thus, the model absorbs knowledge about actual cases and uses it to predict diseases.

After training, the model, is evaluated for performance and effectiveness. Including such parameters as accuracy, completeness, and F1 measure.

Incorporating the model in medical practice is the next stage of the process. As for Cifromed, the model is employed to estimate the probability of getting chronic diseases, including diabetes, cardiac diseases, and cancer. The model is linked to the electronic medical record systems to update the data and deliver instant prognoses. This enables the doctors to make better recommendations and prevent diseases. For instance, the model can determine that a patient is likely to get diabetes through their medical records and genetic data and, therefore, the doctor can advise on lifestyle changes.

The application of LLM also plays a part in enhancing the standard of health care in Cifromed. The model processes patient information and determines possible complications, enabling doctors to give better advice and treatment. This results in the early detection of diseases and boosts the health status of patients. Cifromed stated that LLM has increased diagnostic accuracy by 18%.

LLM is applied to interpretation of laboratory findings and handling of medications. Due to the application of LLM, processing of laboratory results has been done in a quarter of the initial time.

The Cifromed LLM enhanced confidentiality data protection. The model processes data in a de-identified way to avoid revealing the patient identities.

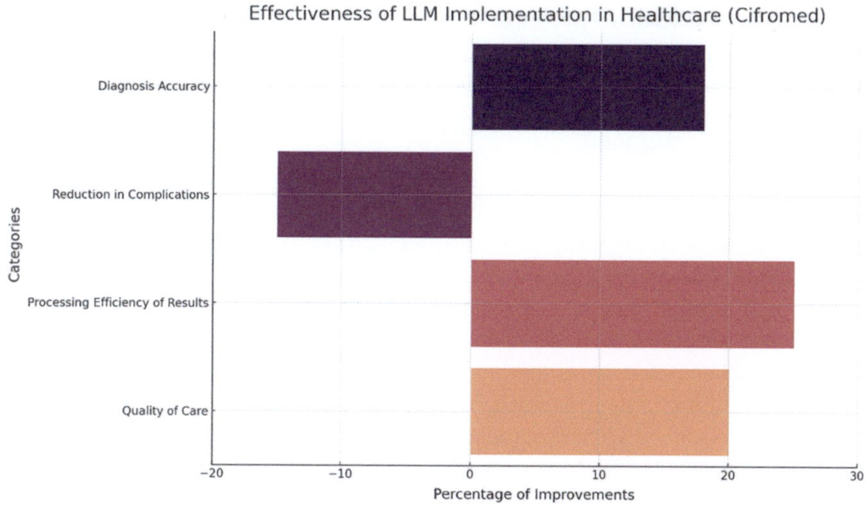

Fig. 4.4 LLM implementation in Cifromed

Frost and Sullivan estimate that applying LLM in medicine can increase the level of diagnostic precision by 15–20% and decrease the expenditure of treatment by 10–15%.

This is as a result of more and more accurate predictions of diseases being made and better management of risks. The LLM, therefore, makes it easier for healthcare providers to understand the status of their patients and thereby, be in a better position to provide better and more specific health care services.

LLM implementation at Cifromed, improved the performance of the organization, provides better diagnoses, and boosted the quality of services delivered.

Figure 4.4 depicts the changes in the healthcare delivery system, such as the diagnostic efficacy, decrease in the rate of complications, faster turn-around time in lab results, and enhanced quality of care.

The examples of Sberbank and Cifromed highlight that the implementation of LLMs in different industries has brought a positive change as it has enhanced process accuracy and efficiency and improved the quality-of-service.

The integration of new technologies in investment portfolio management is among the critical factors that define the effectiveness of management companies. The most popular and efficient technologies that are widely applied in this field include fuzzy search and hybrid recommendation systems. These technologies offer better data analysis, thus enabling investors to make better decisions in the investment and management of portfolios.

Soft search is a search technique that enables the identification of matching information irrespective of the fact that the user's query or data contains typos, misspellings, or different formats. This technique finds application especially in cases of analyzing large volumes of textual data like financial reports, news briefs, analytical reports, and social media. Soft search, in the context of portfolio management, proves useful for investment companies as it makes it possible to access important information from different sources within a short time frame, leading to a better analysis of market conditions.

Text to vector conversion is one of the initial steps that has to be taken when using soft search. Today's NLP techniques, like BERT and GPT, convert the textual input into fixed-sized embeddings which can then be used for information retrieval and matching. These vector representations maintain the semantics of the words and provide high accuracy and reliability of the search outcomes.

Applying the soft search is possible through tools like Elasticsearch and NLP models. Elasticsearch enables the indexing and searching of data with the help of efficient algorithms for large data sets. When combined with vector representations of text developed from NLP models, Elasticsearch is able to query text in complex ways and produce useful results even when the input text contains errors or is similar in wording. This improves the quality of the data analysis, making investment decisions far more effective.

One of the other important technologies applied in investment portfolio management is hybrid recommendation systems. These systems contain both collaborative filtering and content analysis to make recommendations on investments. Collaborative filtering is based on the idea that the recommendation system should follow the behavior of users; therefore, it can recommend investment strategies that match a user's profile. For instance, if a set of users with the same investment targets and histories receive good results, the system can suggest similar investments to users with similar characteristics.

While the above-discussed approaches analyze quantitative data, content analysis, on the other hand, analyzes textual data from financial reports, news and analyzes, to determine the potential of different investment instruments. NLP models are employed by the content analysis systems to determine important features about the text and its possible impact. For instance, a news article may report a certain event connected with a specific firm; the recommendation system can then evaluate the effect of this event on the company's stock value and use the information in any subsequent recommendations.

Hybrid recommendation systems incorporate both of these types in order to give the users better and more relevant recommendations. This way, they can consider a large number of factors that can affect users' investment preferences and complement the analysis of user behavior with textual data analysis. This improves the accuracy of recommendations and assists management companies to develop their portfolio without exposing themselves to too much risk, in order to return more profit.

Examples of this are ensemble learning techniques like Gradient Boosting or Random Forests, where two or more models or analysis procedures are used to generate a single output. These systems can also link with investment management software, which will automatically update the recommendations as and when new information and data come in, as well as when the market undergoes any change.

Hybrid recommendation systems are useful in investment portfolio management because they can be used in tasks such as risk evaluation, predicting the future performance of investments, portfolio rebalancing, and creating an individual's investment plan. These systems can also include the specific preferences of the investor, for instance risk level, time horizon and target returns.

Combining hybrid recommendation systems with other techniques like soft search and text-to-vector representation, it is possible to get effective and versatile tools for analyzing and controlling investments. These technologies give a better and more detailed insight into data to aid in timely and more effective investment decisions.

Besides soft search and hybrid recommendation systems there are other important technologies that can find their application in investment portfolio management. One of such technologies is sentiment analysis, which helps in assessing the sentiment of the market and the probable price trends from the text data. Another important technique is reinforcement learning (RL) which is applied for trading strategies' improvement.

Sentiment analysis refers to the analysis of texts for the purpose of investment management in view of the affective components of texts, including news, social media, analysts' reports and comments.

With the help of state-of-the-art NLP methods, including RoBERTa and XLNet, sentiment analysis can effectively categorize text as positive, negative, or neutral. For instance, suppose a company is discussed frequently in the media with adverse remarks; this may indicate that it has certain issues and, therefore, its stock price may be likely to fall. With the help of such data, management companies can see the tendencies, respond to the changes and adjust the investment strategies.

Combining sentiment analysis with other methods of data analysis helps to get a better view of the market. It allows you to include the qualitative aspects of analysis, which implies financial ratios and the mood of the market. Therefore, through sentiment analysis, one can get to know the sentiment of the market, trends, and possible risks, which are vital in making decisions concerning investments.

Reinforcement learning (RL) is an approach to develop and optimize trading strategies. It is, nevertheless, a learning method that allows agents to learn from their experiences when interacting with the environment to obtain rewards. This approach is particularly useful in the dynamically changing and risky financial market as it allows for relatively rapid changes in the strategy.

In investment portfolio management, RL is used to create agents capable of learning and improving trading strategies. It is the process of performing various activities in the form of a simulated market environment in which the agents trade, examine the consequences of their actions, and consequently adjust their strategies to achieve the maximum long-term return. Some of these models include deep Q-nets (DQNs) which assist the agent in the right decision-making or the model such as the actor-critic models (A3C) which trains the policy and the value functions.

Another point is that RL is able to learn more as the course progresses, so it is not limited by initial knowledge. Therefore, agents are able to change their strategies when new information comes into the market or when the market environment changes, which helps to improve overall performance. This is particularly important during periods of high uncertainty in the market environment, when the existing frameworks and assumptions may become irrelevant.

Other areas of RL in investment management include ensembles, which are a set of several RL agents that can be applied to enhance the potential for making accurate estimates. Such techniques help to formulate various views and approaches to managing the risks and potential returns of the portfolios. For instance, various agents can be specialized in different markets such as equities, bonds or commodities and then the outputs of the agents are combined to generate efficient and consistent investment strategies.

By combining RL with other technologies, including sentiment analysis and soft search, it is possible to create effective investment management systems. Such systems are also capable of collecting, processing, and storing information, designing and optimizing trading mechanisms and strategies, and learning from market changes. In this way, management companies have acquired the means to effectively and precisely manage their portfolios, which leads to high performance and low risk.

The use of RL and SA in investment portfolio management also makes it possible to eliminate certain processes thereby reducing costs. These technologies perform basic functions such as market surveillance and transaction processing, freeing up much-needed time for other critical activities. Automation also removes the human factors prone to error and increases the firm ability to respond to crisis.

The E5 (Embedding from Language Models) model is one of the key models used to transform text data into vectors. This technology is based on the latest deep learning models and therefore allows for the generation of rich feature representations that can be used in numerous applications such as text mining, classification and clustering. In investment portfolio management, the E5 model is used to process the textual data such as the financial statements, news articles and other analytical commentary.

The E5 model is used to convert text into vectors, enabling agents to effectively search for meaning in texts. Vectors derived from the E5 model retain contextual information and relationships between words unlike vectors derived from the word2vec model. This is particularly important for tasks related to soft search and hybrid recommendation systems where it is necessary to achieve high precision and stability of text processing.

The E5 model generates vectors stored in search libraries such as Elasticsearch. These embeddings enable the soft search; the system is able to retrieve documents related to the source text even if there are misspellings or typos. For instance, in news summaries and company reports, the E5 model is useful for data mining and integrating the data with existing data to enable better market trends and risk assessment.

The E5 model can also work with hybrid recommendation systems to improve the recommendation quality. The vectors derived from E5 are used to analyze content data and generate better and more targeted recommendations. In this way, many factors that affect the investment decisions can be taken into account and the quality of portfolio management can be improved. For instance, if the E5 model determines that some texts are positive about a company, the recommendation system can use this knowledge to increase the weight of the company stock in the portfolio.

The E5 model is also used in sentiment analysis. The vectors obtained from it are useful for categorizing texts according to their emotional charge, which helps agents to assess the mood of the market and its potential influence on the price of assets. Because E5 embeddings take into account the tone and context of the text, which are the qualitative characteristics, the accuracy and applicability of the predictions increase when sentiment analysis is used.

It is also advantageous to introduce the E5 model with RL techniques. The vectors derived from E5 can be used as states and actions in the market for RL systems. This results in agents being able to make better decisions strategies. For instance, if the E5 model has predicted that the current market conditions are favorable for certain assets, the RL agent will be in a better position to buy or sell the assets based on the model output.

Sentiment analysis and RL enhance the agent ability to manage investment portfolios. Sentiment analysis enables agents to identify and better predict the feelings and perceptions of the markets and consumers by using sophisticated NLP algorithms to analyze natural language. It helps agents to identify threats and options in the marketplace by arming them with accurate information from news, media, and analyst reports.

RL is used to train agents to learn in a dynamic market environment. The context is a simulated market environment in which agents are trained to learn and refine their trading strategies and are rewarded for doing so. This enables the agents not only to learn improved strategies from historical data, but also to learn from the real-time market dynamics to increase their efficiency and flexibility.

All of the above technologies help agents to analyze data, generate strategies, and make decisions. They help agents reduce risk and improve returns in an ever-changing market environment, paving the way for better investment portfolio management and a more efficient financial sector.

4.2.3 Collecting Data for System Requirements

Raiffeisen Capital is a fairly large company with analytical department of more than a hundred full-time employees. Divided into teams according to areas of expertise. Investment analytics can be divided according to the financial instruments it focuses on: equities, preferred stocks, bonds, futures, options and other financial instruments. It is also divided according to the type of analysis: fundamental analysis (calculation of indicators on the basis of reports, evaluation of the company's strategy, its position on the market, calculation of the stock earnings etc.) and technical analysis (analysis of price quotations, determination of overheated prices etc.). All investment analysts use news. Analysts assessing long-term prospects (fundamental analysis) are interested in news new projects and business development. Technical indicator analysts, on the other hand, focus on the smallest flaws in the market. They take advantage of the short-term news that can change the price. They are interested in current news events and the impact they may have on the price. Those involved in the analysis of debt instruments such as bonds, along with information about the issuer company, pay attention to the changes in interest rates, potential changes in the credit status of an industry etc.

To best engage we propose to conduct the communication in two phases: questionnaires.

The questionnaire should include questions to identify the financial tools and analysis methods, as well as the relevant news types—e.g., corporate, economic, political or social—and the related sources of information available. It is therefore important to identify the tasks of the staff in relation to these news feeds, and the challenges they face. Wish list questions should identify the desired aspects for a digital assistant and the related information. The questionnaire should include both "open" and "closed" questions, the latter to identify the most active employees and select the best respondents.

1. The results of the electronic questionnaire are forwarded to the relevant staff. After the questionnaires have been completed, the results are analyzed, key trends and issues are identified, and a report is produced. The interviews with the executives provide very important information for defining the system requirements.

To explore potential options and features of the digital assistant, the participants are interviewed based on the questionnaire results. The pool includes primary and secondary users and other stakeholders.

The interview plan includes the list of existing processes and tools, the points of issues and concerns, and the requirements for the future digital assistant system including its features and interface. Additional interviews may be required with certain stakeholders to obtain more specific data on their requirements and constraints and/or to provide more detail.

The interviews are recorded to extract key points and conclusions. The data is then cross-examined to determine the common requirements of the key stakeholder groups. These include functional requirements, expected outcomes, and implementation priorities.

4.2.4 Business Requirements

The design principles and patterns of the newsfeed analysis system are listed in a document that specifies the business goals and the high-level data objects. This document provides a foundation for subsequent development phases and helps to ensure common understanding between the stakeholders. It contains the requirements, both functional and non-functional (i.e., quality attributes), needed to build and maintain the system. The pattern of such a document is as follows.

1. General requirements

1.1. Purpose of the System

The system identifies news feeds for market predictions to recommend investment portfolios. It must be compatible with the existing financial systems and provide real-time data.

1.2. User Interface

The interface of the system must be user-friendly even for new users and must include the forms for querying, displaying results and providing feedback.

1.3. Reporting and visualization

The system must report the data in a simple but effective way, including graphs and charts. Data must be visualized to allow interactive and detailed analysis.

2. Functional Requirements

2.1. Data Collection

The system must extract data from financial news, analyst digests and social media. This data must be updateable in real time.

2.2. Validating the Queries

The system must validate the user queries and verifies that all the data required is specified. For a wrong query, the system must indicate the errors made by the user.

2.3. Text to Vector

The system must employ the NLP techniques in order to convert the texts into vectors that allow for highly efficient analysis results.

2.4. Soft Search

The system must use soft search in the text vectors to match the relevant fragments, including source text misspellings.

2.5. Sentiment analysis

The system must determine the likely reader reaction, such as positive, negative, or neutral, regarding the market condition and its effects on the asset price.

2.6. LLM-Based Response generation

The system must provide textual answers through LLM-based data analysis. The answers must be appropriate and helpful, i.e., provide correct and useful information.

2.7. Evaluating User Satisfaction

The system must rate the level of user satisfaction from the responses and store the ratings in a database for future use.

3. Non-Functional Requirements

3.1. Performance

The system must process multiple concurrent queries in a few seconds, ensure efficient data processing, and support multiple users and large data sets.

3.2. Scalability

The system infrastructure should be horizontally scalable, i.e., expandable in terms of the number of hosting servers.

3.3. Reliability and availability

The system must be highly reliable and available with minimal downtime and maximum uptime and automatically recoverable from critical failures.

3.4. Security and Privacy

The system must protect data through encryption and user authentication. Access to data must be restricted according to the organization's security and privacy standards.

3.5. Integration with External Systems

The system must be connected to financial services such as stock exchanges and news aggregators. An open API should provide compatibility with third-party software and services for proper integration.

4. Development and maintenance

4.1. Documentation

The system architecture document must provide information on the structure, elements and their relationships for key stakeholders including developers, administrators, and end-users.

4.2. Maintenance and Support

The system must be maintained and updated to ensure its proper functioning according to the policies that ensure its maintainability and up to dateness.

4.3. Feedback and improvements

The system must collect user feedback to ensure its quality and improve its structure and behavior.

BPMN visualization

First, the user makes a request to start the news data analysis. Such a request may include details of specific tickers and using a certain timeframe and/or source(s) of data. The system receives and validates the request by verifying the data for the required fields. If the request is invalid, the system notifies the user, who updates the input and resends the request (see Fig. 4.5).

If the request is valid, the system saves it to the database. It monitors the status of the request, updates the user and sends the request to the system module that converts the text into vectors for fast processing of large text data.

The converted vectors are then forwarded to Elasticsearch and the soft search mechanism finds the text fragments associated with the ID. Due to the vector embedding, ElasticSearch will process even misspelled text correctly (see Fig. 4.6).

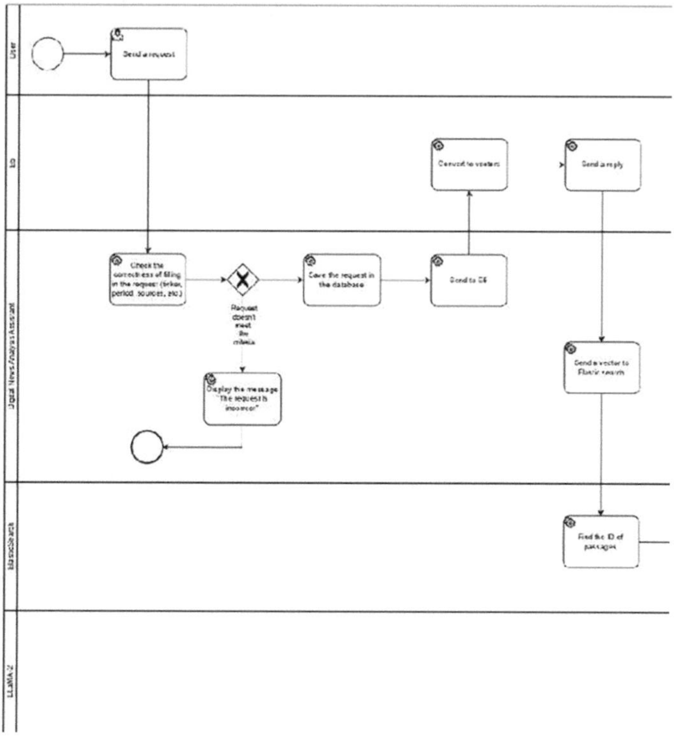

Fig. 4.5 User request validation for news data analysis

After the search, the system checks the confidence level of the extracted data by matching the retrieved fragments to the query and the context. If confidence is low, the user is notified and directed to another analysis method or search criteria. Otherwise, the text is stored in the database and passed to the LLM module. The LLM module generates a response based on the data.

Following the response, the system then asks for a satisfaction rating. The user can also rate the response quality, and the rating is stored in the database to be used for future reference, feedback, and learning (see Fig. 4.7).

The process is distributed across system components, including the E5, ElasticSearch, and LLM modules. First, the user submits a query, which is checked by the system and converted into vector form by the E5 module. ElasticSearch then searches these vectors for similar text fragments. Further, if there is sufficient confidence, the text fragments are passed to the LLM module to form a response. The text is stored for the user to view and rate.

After defining the business requirements and explaining the BPMN processed us, we identify the system features and key quality attributes.

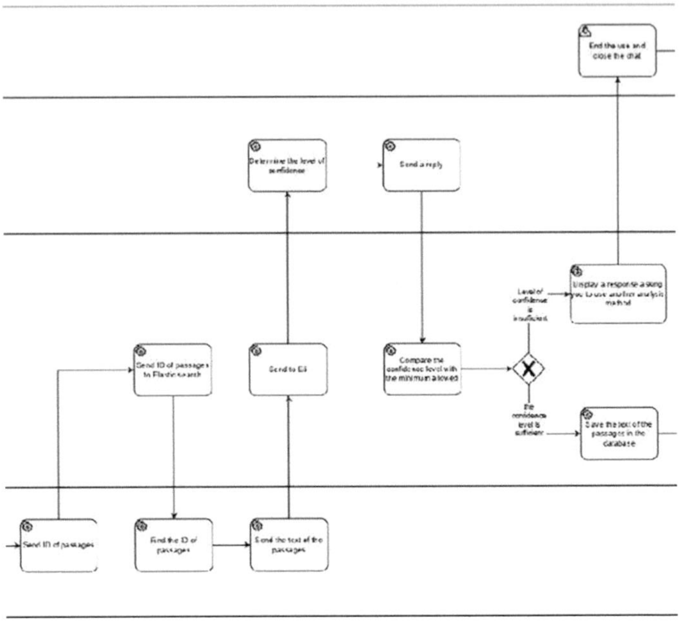

Fig. 4.6 From valid request to vector-based search in Elasticsearch

Expected Performance

1. High speed query processing. User requests should be processed by the system within a few seconds. The total processing time includes the collection and analysis of heterogeneous data, and response generation. Performance enhancing factors include high-performance servers, optimized algorithms, and parallel computing
2. Multi-user mode. The architecture must effectively control and allocate computing resources and isolate individual user requests. To handle more concurrent users, multiuser mode enables horizontal scaling
3. Low response time. Includes the following techniques: caching, data pre-processing, and reduced data transfer between system components. Latency is reduced through efficient data transfer protocols and optimized network configuration
4. High availability and reliability. This includes a robust design with redundancy for the essential elements, system clones to take over in the event of a failure, and data backup. System reliability is ensured by regular maintenance checks and other preventive measures
5. Efficient handling of large volumes of data from multiple sources through powerful GPUs and distributed computing frameworks. Optimized data processing algorithms and parallel computing help reduce analysis time, improve prediction accuracy, and expand the system to support big data volumes.

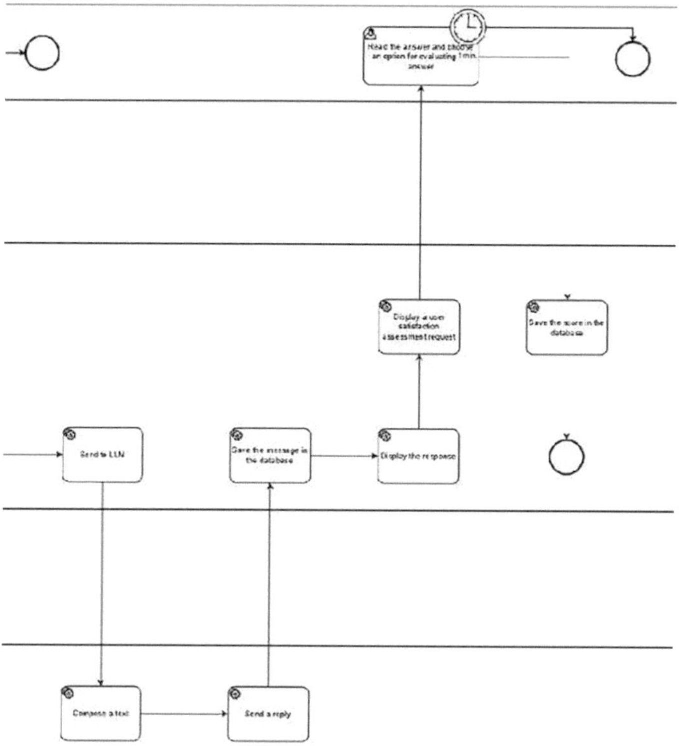

Fig. 4.7 LLM-based response generation and satisfaction rating

Data processing features

1. The system must collect data from multiple sources: financial news, analyst statements, social networks, and other platforms. Diverse data sources ensure a rich dataset for proper analysis of market trends. Data must be collected in (near) real time, especially for news feeds
2. For text analysis, the system uses recent advances in the NLP models such BERT & GPT digitize the text and include vector conversion to perform soft search, which supports incorrect queries or misspelled words
3. Sentiment analysis is based on the emotional tone of the texts such as positive, negative and neutral, which allows qualitative assessment of market trends
4. Data processing incorporates ML applications to identify latent behaviors and characteristics. Although the system is purely deterministic, these algorithms help to learn from the system data and user feedback to improve prediction accuracy
5. Data security and privacy are ensured through encryption and authentication services. Data access is based on certain security and privacy standards and regular security checks to prevent data compromise

6. Regularly updated models and algorithms ensure that the system delivers accurate, efficient results and innovative technology direction.

Integration with external systems

1. Connection to financial platforms and news aggregators. The system must be connected to financial websites and news aggregators including stock exchanges, news agencies/services, and social networks. This ensures diverse and accurate data, which is essential for valid analysis and prediction
2. APIs for external system integration. These APIs must be properly documented, secure and interoperable to reduce the data retrieval time
3. Flexible integration. The system must be adaptable and extensible to cope with large volumes of data and multiple external systems. The system must be scalable and flexible using cloud technology and microservice architecture
4. Data processing and normalization. To successfully integrate with systems that are heterogeneous in format and structure, the system should be able to adjust and standardize the data to prepare it for analysis and cleansing. This involves removing duplicates, verifying and correcting errors, and ensuring consistency
5. Integration monitoring and management. This includes validation of external data sources and data access and processing. The system must detect and correct integration errors such as data source unavailability or data transfer failure, and alert administrators
6. Data security. External system integration must meet stringent security and privacy standards. This includes secure communication protocols such as HTTPS and SSL/TLS, as well as user/service authentication and authorization. Data access must be restricted and monitored, and integration activities must comply with regulatory standards and corporate policies
7. Integration process assessment and improvement. The system must use feedback from users/administrators to improve the integration procedures. This includes feedback assessment, defect detection and management.

For efficiency and reliability, system functionality and quality must be assessed in terms of performance, data processing, integration points, data security, user privacy, real-time processing etc.

Multi-source data acquisition, vectorization and sentiment analysis using sophisticated NLP techniques enables subtle trends and patterns to be uncovered, improving quality of analysis and decision-making.

System integration enables versatile data through open, scalable APIs that easily accommodate changing requirements and boost performance in the heterogeneous environment. Data processing and normalization, integration control and management, and data protection provide a solid framework for managing multi-data sources.

The above pattern supports high system efficiency for accurate and timely investment recommendations. The system meets user expectations for key features and qualities such as performance, data accuracy and consistency, making it a valuable investment portfolio management tool. Well-designed architecture and interoperability enhance the quality of the system and make it resilient to crisis, which increases its market value.

4.3 Results and Conclusion

The development of the newsfeed analysis system to identify market trends and control investment portfolios involved a number of stages, from requirements management to the architecture of the interaction with external systems. Key system qualities identified included performance, reliability, and usability.

System performance depended on such aspects as response time in query processing, scalability, availability and reliability. Response times of a few seconds were achieved by improving processes such as data acquisition, pre-processing, analysis, and response generation based on parallel and high-performance computing.

Scalability, availability, and reliability have been achieved through an API-based flexible, fault-tolerant and robust system architecture.

Another critical system attribute, predictive accuracy, was based on state-of-the-art NLP models, sentiment analysis, and other techniques, including text mining and vectorization. This attribute supports efficient decision-making that increases profitability and leads to crisis resilience by reducing risk.

A number of activities, such as data collection, processing and analysis were automated. This allowed the end-users to focus on investment management strategies and increased efficiency and user satisfaction. The other factor that increased the accuracy and credibility of the system was real-time connectivity to external applications. Other quality attributes included data security, confidentiality, and privacy.

Finally, let us summarize the business-critical system metrics that demonstrate its quality. The average query response time is typically within 2–3 s for 10,000 complex queries per day, availability is 99.9%, and typical prediction accuracy is 85–90%. The results demonstrate improved system throughput and reliability, user satisfaction, and prediction accuracy.

References

1. Guerard, J.B. (ed.): Handbook of portfolio construction: Contemporary applications of Markowitz techniques. Springer New York, NY (2010). https://doi.org/10.1007/978-0-387-77439-8. ISBN (Hardcover): 978-0-387-77438-1. ISBN (Softcover): 978-1-4899-8302-2. Copyright: Springer-Verlag US 2010
2. Sönksen, J.: Machine learning for asset pricing. In: Chan, F., Mátyás, L. (eds), Econometrics with machine learning, vol. 53, pp. 337–36. Springer, Cham (2022). https://doi.org/10.1007/978-3-031-15149-1_10. ISBN (Print): 978–3–031–15148–4. ISBN (Online): 978–3–031–15149–1

3. Cernauskas, D., Tarantino, A.: Risk management in finance: Six sigma and other next-generation techniques. Wiley, (2015). ISBN: 978–1–119–19781–2
4. Brugière.: Quantitative portfolio management, 1st ed. Springer, (2020). ISBN-10: 3030377393. ISBN-13: 978–3030377397
5. Montier, J.: Value investing: tools and techniques for intelligent investment, 1st ed. Wiley, (2009). ISBN-10: 0470683597. ISBN-13: 978–0470683590
6. Vaswani, A., Shazeer, N., et. al.: Attention is all you need. In: Proceedings 31st Conference on Neural Information Processing Systems (NIPS 2017), Long Beach, CA, USA

Chapter 5
Supporting the Digital: IT Process Improvement for Service Company

Abstract This chapter focuses on IT transformation in diverse service environments, their common tradeoffs being high competitiveness, rapid dynamics, and processes concurrency. In the above landscapes, process complexity, tightly coupled with multi-dimensional diversity, is the primary crisis source. This chapter presents the agile way of IT transformation based on the Plan-Do-Check-Adjust lifecycle. Other ingredients elaborating this high-level pattern include ITIL and ITSM process frameworks, the BPMN, UML, and IDEF standards, and the supporting Draw.io toolkit. The above pattern was applied to produce process optimization patterns, for service management organizations to meet the new challenging conditions of fully digital IT services. The optimization pattern suggested essentially reduces servicing time and promotes customer satisfaction.

Keywords IT service management · Process improvement · Incident management · IT infrastructure library

5.1 Introduction

In a constantly changing environment with the dynamic development of the IT, all companies need to strengthen their positions in the market to ensure financial stability. Companies needs to be responsive and continuously improve. Regular analysis and optimization of business processes helps to achieve these goals.

There are many ways to optimize business processes: conducting a SWOT analysis (analysis of weaknesses and strengths), using cause-and-effect diagrams benchmarking, following Lean methodology, and others. One of the most practical approaches, which is used in many successful companies, is the ITIL/ITSM methodology. This is one of the most versatile and practical optimization methods [1].

S. V. Zykov, *IT Crisisology Patterns and Practices*, Smart Innovation, Systems and Technologies 449, https://doi.org/10.1007/978-981-95-1260-7_5

5.2 Theoretical Background

5.2.1 Business Process

The term "process", like many English words, is of Latin origin. Process, translated from Latin, means movement forward, advancement.

Any business consists of many processes, the quality and efficiency of which directly affects the development and profitability of the company.

Among many definitions of the concept of "business process", we selected one by Eliferov and Repin, their book "Business Processes. Regulation and Management" [2].

From "*A business process (BP)* is an ordered set of activities aimed at transforming inputs into outputs in order to satisfy the consumer." The main task of business process analysis is to improve the efficiency of organization management [3].

There are several classifications of business processes. Let us consider the main and best known [4]:

Association of Business Process Management Professionals, in thein Common Body of Knowledge, identified three types of business processes [5]:

(1) *Core processes* are cross-functional processes, the main goal and result of which is the direct creation of value for the consumer. It is these processes that serve as the basis for the value chain
(2) *Ancillary processes* are necessary to maintain the core processes, usually through the management of the resources and/or infrastructure required for the core processes. The main difference between the primary and secondary processes is that the secondary processes do not provide any direct value to the end user
(3) *Management processes* are to measure and control the company activities. They ensure that the main and auxiliary processes are properly designed and executed in accordance with the established operational and financial objectives, as well as regulatory and legal constraints.

There are also two main *Russian classifications* of business processes [6].

In Relation to the client:

a. Basic (value-added)
b. auxiliary (adding costs).

In relation to the value chain [7].

(a) Main business processes (generate company revenue)
(b) supporting business processes (provide support for the company's infrastructure)
(c) business development processes (improve the company)
(d) management business processes (help to manage the company).

To carry out a high-quality reengineering of business processes, it is first necessary to analyze all existing processes, creating a descriptive model of their architecture.

"*Enterprise architecture* is an exhaustive description (model) of all key elements, as well as the relationships between them, including the process of supporting the modification and transformation of enterprise business processes with the help of information technology" [8].

Enterprise architecture consists of the following components:

1. Objectives and intentions
2. Business processes
3. Structure of the organization
4. IT systems
5. Information & resources.

5.2.2 *Optimization of Business Processes*

To optimize and introduce new technologies, a company first needs to describe and analyze all existing business processes.

The main task of reengineering is to identify the business units and optimize them in order to achieve certain goals.

The reengineering process consists of the following main stages:

1. Proposal for an optimization project
2. Preparation of the necessary documents
3. Planning, approval and approval of optimization stages
4. Description of existing processes—"as is"
5. Development of a reporting and work control system
6. Monitoring and control of the implementation of work
7. Correction and improvement of the BP
8. Analysis of the optimization results
9. Completion of the project documenting.

However, after the last stage, the reengineering should not stop.

Instead, it must regularly optimize business processes and monitor changes. To stay crisis-responsive.

Neave provides data [9] that the optimization of any process consists of 4 stages (shown below in Fig. 5.1), which constitute the Deming cycle, PDCA (see Fig. 5.1) that consist of four key steps:

1. Plan
2. Act or Execute (Do)
3. Check
4. Reschedule or Adjust (Act).

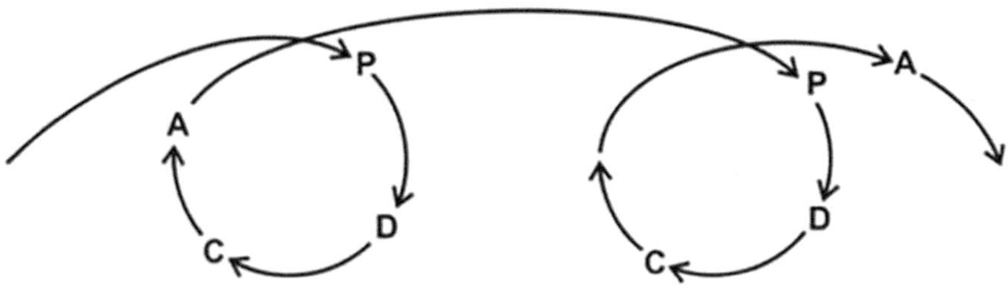

Fig. 5.1 PDCA lifecycle

The PDCA cycle is a repetitive four-step management method to control, permanently improve, and optimize processes and the level of services provided. [10].

The Deming cycle is widely used in BP optimization. With its help, it becomes possible to carry out reengineering until the BP meets certain requirements.

When using the Deming cycle, the optimization process occurs in such a way that it repeatedly goes through all four stages. At each stage, all shortcomings and gaps are identified, and after which are eliminated. The process is improved again (compared to its previous version), and the cycle begins all over again. In this way, the process is brought to the required level of efficiency, after which the company can proceed with its final implementation and start optimizing other processes.

5.2.3 *Business Model*

A business model is a formal description of business processes. It can be presented in the form of a table, text, symbols, graphically, etc. One of the main areas of application of business modeling is the reengineering (or optimization) of business processes.

The main objectives of BP modeling are:

(a) Providing an understanding of the company structure and changes in ongoing processes
(b) Ensuring understanding of current organizational tasks, issues, and opportunities
(c) Control of the of goals and objectives of the company
(d) Forming of requirements for software.

There are several types of methodologies for describing and analyzing business processes. Below are the main ones:

1. Creation of a business process model. The most widely used is the IDEF0 standard
2. Data Flow Modeling. DFD notation helps to display the order of the process, and information flow
3. Workflow modeling. IDEF3 is used to depict workflows.

In practice, the IDEF0 standard is most often used in projects to optimize local business processes in small and medium-sized companies.

In addition to those described above, there are many other methods for modeling business processes. The most popular modeling languages are Business Process Modeling Notation (BPMN), Object-Role Modeling (ORM), and Unified Modeling Language (UML), each of which we discuss in more detail below, along with the IT Infrastructure Library (ITIL).

Object-role modeling (ORM) is used to model semantics. The ORM is often used for data modeling and software development.

The object-role model uses graphical symbols, based on first-order predicate logic and set theory, to create an unambiguous definition of an arbitrary universe. Most often, ORM is used for the purpose of designing databases [12].

Unified Modeling Language (UML) is a graphical description language for modeling objects in the field of software development, system design, for describing and creating models of business processes, displaying the structure and architecture of an organization [13].

Business Process Notation and Model (BPMN) is a notation in XML for modeling business processes [14].

5.2.4 ITIL/ITSM

ITSM guidelines and standards help organizations implement an ITSM strategy. A 2017 report, based on a survey of IT leaders shows that the five most commonly used ITSM approaches are: Information Technology Infrastructure Library (ITIL) (47%), Multi-Level Product Management Business Process Model (eTOM) (36%), Management Challenges for Information and Related Technologies (COBIT) (36%), Microsoft Operations Framework (MOF) (34%), and ISO/IEC 20,000 (29%) [15].

ITIL—"Information Technology Infrastructure Library", is a set of detailed IT service management (ITSM) practices to align IT services with business needs [16].

ITIL describes processes, procedures, tasks, and checklists that are neither organization-specific nor technology-specific, but can be applied by an organization to develop strategy, provide value, and maintain competency.

This allows the organization to establish a baseline from which it can plan, implement, and measure. It is used to demonstrate compliance and measure improvement.

IT Service Management (ITSM) refers to all activities, e.g., (directed policy, organization and structuring of processes, as well as ancillary processes) for designing, providing, using, and controlling the level of service provided to customers [17].

ITSM differs from other approaches in that its main principles are the application of a process management approach, focusing on the interests of customers and the associated service, rather than on the technologies and software.

The main purpose of ITSM is the transition from the traditional model of supporting IT infrastructure to a new one focused on servicing the core business.

However, such a transition can be complicated by the fact that it implies a rather radical revision of the positioning of IT departments in the overall structure of the organization.

The implementation of ITSM is based on the construction of the following processes (see Fig. 5.2):

Let us take a closer look at each of the processes [18].

- The problem management process is based on methodologies to find the root causes of incidents, and develops procedures to prevent the recurrence of similar problems
- The incident management process is about creating approaches to ensure that incidents are resolved quickly
- The configuration management process controls all components of the infrastructure. All important elements are recorded in the configuration management database
- The release management process allows to deploy, integrate, store, and test releases
- The change management process controls of changes to the configurations of IT objects

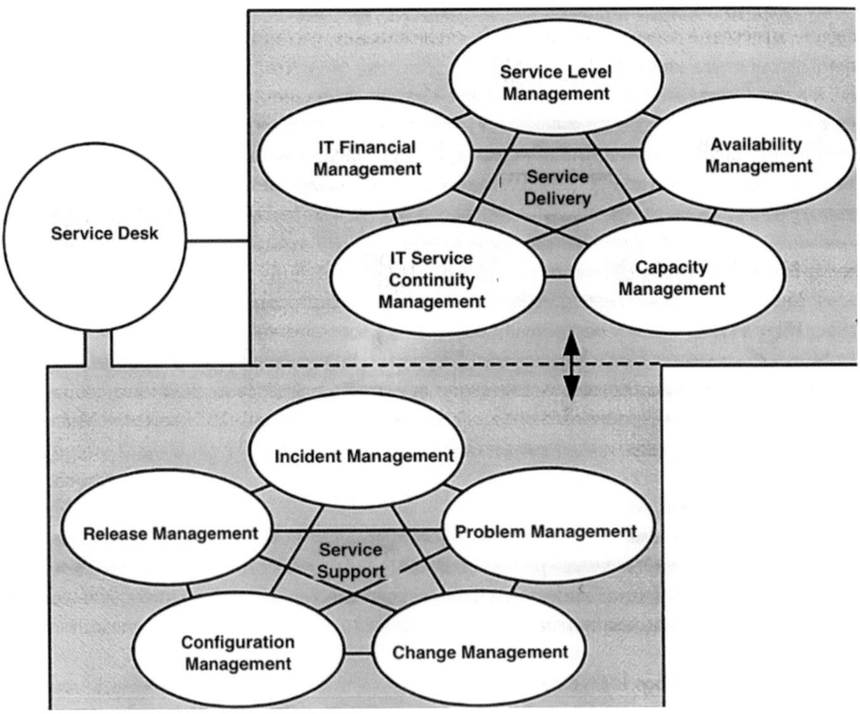

Fig. 5.2 Processes performed by the service desk

- The process of capacity management allows to optimize costs
- The service level management process is designed to ensure the agreements between the customer and the contractor providing IT services. Including service level agreement and key performance indicators
- The continuity management process is concerned with emergency response options and their consequences
- The availability management process ensures the placement of resources and provides technologies and methods services availability
- The financial management process refers to economic issues.

According to these processes, the company sets certain indicators of efficiency and the level of services provided.

5.2.5 Key Processes in a Service-Oriented Company

For the key processes of a service-oriented company, the world-recognized *ABPMP* classification, chapter, will be used to divide the processes into three groups: main, auxiliary, and managerial.

1. *Main processes* (or operational) are cross-functional these directed create value. For a service-oriented company, we will have the following processes:
   - Order acceptance process (phone, online appointment, in-person, etc.)
   - Order processing process (data recording, prioritization of order fulfillment, data transfer to third-party services, etc.)
   - Order fulfillment process
   - The process of creating and designing a set of services provided by the company
   - Marketing processes (advertising, promotions, etc.)
2. *Supporting processes,* required to maintain the main processes, are usually carried out through the management of resources and/or infrastructure required for the main processes:
   - Data processing process (analysis of the target audience, formation of a customer base, etc.)
   - Maintenance (setting up equipment and third-party services, etc.)
   - Administrative processes (cleaning of premises, operation of equipment, etc.)
   - Accounting and budget planning
   - Quality control process (e.g., collecting and analyzing KPI metrics)
   - Recruitment process
3. *Management processes* are to measure and control the company activities. They are necessary to ensure the appropriateness of the main and auxiliary processes in accordance with the goals of the company.

- The process of creating and improving company development strategy (strategic management)
- Long and short-term planning processes
- Processes related to the development of human resources
- Corporate management processes
- Management of business processes.

5.2.6 BPMN Model

Using the BPMN language, let us build models of basic processes of a service-oriented company. The models use the draw.io pattern, (the script is in Appendix 1).

Now let us consider each of the models in more detail.

Let us start with the main process—*the process of receiving and placing an order* (see in Fig. E.1 of Appendix E):

1. Start event: an order has been received from the client
2. Action: check whether the service requested by the client exists in the list of services of the company
3. If yes, action: check whether there are resources (material, labor, etc.) to perform the requested service

   3.1 If not, action: inform the client about the impossibility the service

4. If yes, action: check whether it is possible to book a time in the company and/or personal schedule of the contractor

   4.1. If not, action: inform the client about the impossibility of performing the service

5. If yes, action: inform the client about the possibility of fulfilling the order

   5.1 If not, take action: inform the client about the impossibility of performing the service

6. Action: receive answer
7. Action (regardless of response): Enter data into the database of incoming orders
8. Finish event: the process is completed.

Next, let us consider the *data processing process*: for example, analyzing the prices of various services in the local market model in Fig. E.2 of Appendix E:

1. Start event: a request has been received to analyze data in the market
2. Action: find data
3. Action: Save data to the database
4. Action: Process data manually
5. Action: Generate a general report on the received data
6. Action: Provide a report
7. Finish event: event process completed.

Again, this process is based on the example of a company using the services of an analyst who, manually or with the help of auxiliary software, conducts a study to identify the average prices in the service market in a certain region, which can often be inconvenient because it will take a lot of time, inaccurate because the analyst may miss some data or use irrelevant data, and there may be an error in calculations, etc.

The incident management process is one of the key functions.

An incident is an unplanned interruption of a service or a decrease in the quality of a service (ITIL V4).

The incident management process is presented in more detail below (see Fig. E.3 of Appendix E).

1. Start Event: an incident occurred
2. Customer Action: Contact technical support
3. Technical Support Action: Log a case
4. Client's action: Describe the incident, its scenario
5. Categorize Case Service Request and provide initial support
6. If yes, process the service procedure request and complete it
   6.1 If not, compare it with familiar problems and methods of solution; check if it matches
   6.2 If yes, resolve the incident and restore
      6.2.1 If not, conduct an investigation and complete a diagnosis
7. Notify the customer to fulfill the request for incident elimination
8. Finish Event: Incident completion.

The process can also be streamlined by creating a system for recording incidents and service requests.

5.2.7 Model Using ITIL Technologies

Let us start with optimizing *the process of receiving and forming an order*.

There are two main options: creating your own online scheduling system or implementing existing systems.

The first option is more advantageous in the long term, and more suitable for medium and large businesses, since it requires a lot of resources for development, of the system.

The second option is to use a third-party online scheduling system, which is more suitable for small and medium-sized businesses in the short term, since requires less time on development.

However, regardless of the chosen system, they will have approximately the same functionality and purpose. This will simplify communication between the client and the company, since in order to receive all the necessary data, the client will need to log in to the system and select the appropriate time and service. The main advantage of the online system is its availability.

An optimized model of the process of forming an order follows (see Fig. E.4 of Appendix E):

1. Start event: an order has been received from the client
2. Action: check the correctness of the filled data. Is the data correct?
3. If yes, confirm the order
   3.1. If not, inform the client about the impossibility or incorrectness
4. Finish event: the process is completed.

Next, let us consider *the process of data processing*. In many companies, the process of processing and research is carried out by an analyst. Often, the search and processing of data takes up to 80% of the total work, which, of course, affects both the productivity of a particular employee and the efficiency of the entire company.

To solve this problem, it is proposed to introduce software that will collect and consolidate data—i.e., engage in such a process as data mining, analysis.

Data mining is the process of discovering previously unknown, non-trivial, and practically useful knowledge in raw data that is necessary for decision-making [19].

Here, as with the optimization of the process of receiving and forming an order, there are similar ways to solve it.

An example of such a solution is the simplest parser.

A parser is a program, service, or script that collects data from certain web resources, analyzes it, and creates a report in the requested format.

Now, let us take a closer look at how such a system will help optimize the data processing process, (see Fig. E.5 of Appendix E).

After optimization, the data processing process consists of three steps (previously five):

1. Start event: a request has been received to analyze data in the market
2. Action: Process data with software
3. Action: Make adjustments or supplement with data obtained
4. Action: Provide report
5. Finish Event: Process of has been completed.

Consequently, actions such as searching for data, entering it into the database, processing it manually, as well as generating a report from scratch, are no longer necessary since the software will do all of this.

Before the analyst comes to study the average prices in the market, the software will have already collected the necessary data in the database over a certain period of time (for example, a month), analyzed it, reported. The analyst will only have to check, correct, and add any missing information, if necessary.

The final example of a process that will be optimized is *the incident management process by* Service Desk system.

Service Desk is the core IT function within the IT Service Management (ITSM) discipline defined by ITIL. It is designed to provide a single point of contact (SPOC) users and IT staff [20].

The ITIL approach considers the Service Desk as a central point of contact between service providers and customers on a daily basis [21].

Next, let us look at what the improved incident management process will look like, (see Fig. E.6 of Appendix E).

After the process of optimization, we get two actions instead of six:

1. Start Event: an Incident occurred
2. Action: Register the incident (can be performed automatically). Service request?
3. If yes, process the service request and complete it
    - 3.1. If not, match them to known patterns methods of solution; check if it matches
    - 3.2. If so, resolve the incident and restore
        - 3.2.1. If not, conduct an investigation and diagnosis
4. Finish Event: Incident complete.

When making contact, the client describes the incident by filling out a form, selects the type of problem from the list, which then allows you to classify the request and send it to the specialist dealing with this type of problem.

The program then registers the incident, sends a notification to a Service Desk employee, who, in turn, processes the request. After that, the system marks the request as completed. The customer is then notified by the system that the requested service has been completed and can resume work.

If the case is an incident, the Service Desk employee will check whether the it can be compared with familiar cases. If one or more of the existing solutions are suitable for eliminating the incident that has arisen, it directly solves the incident and restores work. Which, as in the case of a service request, it marks the request as being completed and the system notifies the customer.

If the Service Desk specialist does not find such precedents in the history of the company, he then conducts an investigation, collecting and consolidating information on this incident from external sources, and/or performing diagnostics in order to eliminate the incident.

Optimizing this process saves time by omitting redundant actions. The implementation of the software simplifies the work of the client and Service Desk and increases efficiency of the company.

Analysis of the work done

The process of placing an order has been reduced in the number of actions three-fold: from six to two [22].

According to the statistics of the online scheduling platform Yclients, approximately 25% of appointments are made working hours or situations where customers cannot make a call [23]. That is, a company that does not offer online appointment booking loses about a quarter of potential customers.

As for the *information processing process*, the introduction of a parser to simplify the work of analytics will reduce the analyst's work time by several times.

Optimization of this process will allow to conduct analyzes more efficiently minimizing human error, and in a shorter time.

By implementing a Service Desk system, the company reduces the time spent on incident classification reduces and help to effectively distribute tasks.

The Implementation will help to avoid the loss of requests reduce downtime and allow to organize the collection of high-quality statistics on the occurrence of incidents.

According to Okdesk, a service that provides ready-made solutions for Service Desk, the implementation of such a system can reduce the time for resolving requests by 30%, and customer satisfaction increases to 99.5% [24].

5.3 Conclusion

In this chapter, models of several business processes of a service-oriented company were developed and recommendations for their optimization based on the ITIL approach have been given.

It is recommended to optimize a company business processes by implementing software that simplifies the work of employees and customers.

The use of optimized models reduces a company costs, increases efficiency, improves the quality of services provided and the level of service, and communication.

The pattern obtained are of practical value and can be used by a service-oriented company to optimize business processes.

References

1. Jeston, J., Nelis, J.: Business process management: practical guidelines to successful implementations, 1st edn. Butterworth-Heinemann, an imprint of Elsevier, Oxford, UK (2006)
2. Weske, M.: Business process management: Concepts, languages, architectures. Springer, Berlin, Heidelberg, New York (2007). ISBN 978–3–540–73521–2
3. van der Aalst, W.M.P.: Process mining: Discovery, conformance and enhancement of business processes. Springer Verlag, Berlin, Heidelberg (2011). ISBN 978–3–642–19344–6
4. von Brocke, J., Mendling, J.: Business process management cases: Digital innovation and business transformation in practice, p. 2018. Springer, Cham (2018)
5. Davis, R.: Business process modelling with ARIS: A practical guide, p. 2001. Springer, Berlin, Heidelberg (2001)
6. Gerke, K., Petruch, K., Tamm, G.: Optimization of service delivery through continual process improvement: A case study. ISSS/BPSC, (2010)
7. Eliferov, V.G., Repin, V.V.: Business processes. Regulation and management, (2017). INFRA-M, p. 320
8. Business processes and process approach: How to overcome the consequences of consulting marketing. Organizational Structure. http://orgstructura.ru/business-processes-andprocess-approach (Accessed: 01.05.2025)

9. Ryzhko, A.L.: Causal classification of business processes of the enterprise. Manag. Sci., 8(1), 90–99 (2018)
10. Belaichuk, A.A., Eliferov, V.G., (eds.): Body of Knowledge on Business Process Management (BPM CBOK 3.0). Moscow, APUBP, (2015)
11. Repin, V., Eliferov, V.: Process approach to management. Business process modelling, p. 544. Mann, Ivanov & Ferber, (2013)
12. Mamaderov, R.K.: Basic concepts and classification of business processes. № 11–3 (14), pp.79–80 (2016)
13. Krasnov, S.V., Diyazitdinova, A.R.: The concept of the enterprise architecture support system. In: Vestnik Volzhskogo universiteta im. V.N. Tatishcheva No. 2 (19), pp. 60–65 (2012)
14. Astrakhanskaya, S.P.: Russian Universities program, p. 344. Prostranstvo Doktora Deming: Moscow, MGIET (TU), (1996)
15. What Is the Plan-Do-Check-Act (PDCA) Cycle? https://asq.org/quality-resources/pdca-cycle (Accessed: 01.05.2025)
16. Optimization of Business Processes. Beta-soft. http://betasoft.ru/development/business.php (Accessed: 01.05.2025)
17. Krogsie, J., Opdahl, A.L., Brinkkemper, S.: Conceptual modelling in information systems engineering, p. 346. Springer, (2007)
18. Booch, G., Rumbaugh, J., Jacobson, I.: Unified modelling language user guide, 2nd ed., p. 496. Addison-Wesley, (2005)
19. Business Process Model and Notation. Object Management Group. https://www.omg.org/spec/BPMN/2.0.2/ (Accessed: 01.05.2025)
20. Ruiz, M., Moreno, J., Dorronsoro, B., Rodriguez, D.: Using simulation-based optimization in the context of IT service management change process. Decision Support Systems, pp. 1–34. ResearchGate, (2018)
21. What is ITIL? Your guide to the IT Infrastructure Library. https://www.cio.com/article/2439501/infrastructure-it-infrastructure-library-itil-definition-and-solutions.html (Accessed: 01.05.2025)
22. Brenner, M., Garschhammer, M., Hegering H.-G.: When infrastructure management just won't do-the trend towards organizational IT service management//Managing Development and Application of Digital Technologies: Research Insights in the Munich Center for Digital Technology & Management, pp. 131–146 (2006)
23. Matveev, A.Y.: IT management as a basis for improving the efficiency of the enterprise. Int. Organ. Res. J., (1), 39–43 (2016)
24. Khorev O.E.: Analytics and use of quantitative data analysis. In: Proc CAI-2016, UMC UPI, 2016, pp.199–202 (2016)

Chapter 6
Applying Microservices: Data Integration and Graphical Interface Testing

Abstract This chapter outlines a digitalization pattern for data integration in complex environments. The idea is applying the extract-transform-load (ETL) type of transformation to heterogeneous data warehouses and data handling tools. In these large-scale warehouses, data size, dynamics, heterogeneity, and diversity, taken together, constitute a dramatic crisis trigger. To facilitate the above-mentioned compatibility, a layer-based architectural pattern is suggested. To provide the required level of flexibility, this common pattern features metadata modeling facilities and management instruments. This service-oriented methodology also includes microservices, publish–subscribe event handling, and open-source API to facilitate loosely coupled module interaction, which is essential for crisis-resilient data management. Data orchestration and choreography are suggested as possible ways to conquer heterogeneity. The intelligent toolkit used includes Kubernetes, Apache Airflow, and a machine learning-based recommending mechanism to support versatile data and social network integration.

6.1 Introduction

The project, from a technical point of view, is a tool that provides a combination of developed solutions and services.

Let us imagine a real task faced by many teams around the world: "We need to take data from many sources, perform certain transformations and get a beautiful visualization as the end result." Let us consider, an approximate flow of processes, and describe tasks and communications in detail.

- Business User has a request such as "I want the charts to nicely show the profits for each month, over a three year period, for the two main departments". This is what happens most of the time, and now the task of the team is to figure out what exactly, the customer requirements
- Project Manager (PM) is usually the first person to feel the "pressure" from the user. His or her main task is to transform the desires of the business user in to a

S. V. Zykov, *IT Crisisology Patterns and Practices*, Smart Innovation, Systems and Technologies 449, https://doi.org/10.1007/978-981-95-1260-7_6

technical document. As a result, documentation is produced containing clarification for the business user (What to load? Where to load? How to load? Why to load? etc.)

- Further, the information goes to the Data Analyst, who, updates and transfers the received new document to the data engineer
- The Data Engineer creates an ETL pipeline for transferring data from multiple sources to the final structures created by the data analyst reengineer. The team receives pipelines and data ready for further work. The data engineer is also responsible for the visual representation of the data
- Further, Senior Data Engineer, Technical Lead or Data Architect, verify the solution.

The process implies the use of tools, such as Microsoft Word, Git, Airflow, Redash, and IDEs like VSCode.

Wildberries experience allowed to rethink the issue of usability with multiple tools separately. Indeed, very often we, as developers, engineers, data analysts and architects, lose a lot of time on the useless movement between one instrument to another. This included many orchestration solutions—Oracle ODI [7], Airflow [8], Airbyte [9], many databases—Oracle [10], PostgreSQL [11], Clickhouse [12], and the client wanted to reduce the time wasted when they needed to access each of them separately. For convenience in data orchestration, features combine ready-made solutions and integrate them all in one shell.

Microservice Architecture pattern was taken as a basis. API-based microservices architecture fits the distributed structure of the application [5], which allows linking various atomic actions together through the API layer. Message broker, integrated and centralized messaging in the microservices architecture [5].

Integration engines are responsible for obtaining data from sources. They can locate both on servers and on clients. This architecture will not allow unauthorized access to the data and make the system extensible. Message broker in this architecture works as a queue, and integration engine.

Ready-made algorithm integrates data by a GitLab CI Runner [6].

Database Proxy solves security issues loads data into the storage (append, merge upload options), checks the quality of data at the loading stage, etc. and connects databases multiple departments and companies.

The main components also include a metadata repository. To store settings for databases and integrations, descriptions of tables and columns. Dedicated storage will reliably and securely store metadata, and immediately grant access to it when requested.

The heart of the system is the API layer, which deals with user authorization, access control and distribution of tasks. Depending on the access level, it should allow users to perform various actions with the dataset: reading, launching, editing and creating.

The resulting application architecture will give the application ease of development maneuverability, scalability and deployment [2].

The goal of this project is the creation of a software solution that allows to effectively, quickly and conveniently use many existing tools on the market and not waste time on their connection. The creation of this application is planned in the form of a web service, with a clear and intuitive interface, and in future there are plans to expand it to a mobile version. The most important requirement for it is to ensure the connection of many disparate and uncorrelated programs into a single whole.

The "DataTool" project is a group project, with a clear division into roles and developed functionality, and a description of the solution from two sides. Before starting the description of the theoretical part of the study, let us take a look at the solutions of competitors.

6.1.1 Analysis of Potential Competitors

Similar to the concept of DWH, when all the structured data of a company is collected in one place (not to be confused with semi- and unstructured data for which there are the concepts of Data Lake and Data Swamp), a tool is needed that collects all structured data about data (metadata) in one place. Theoretically, all metadata can be model-agnostic by splitting into Application Context (semantics and description), Behavior (how the data is received and how it is used), and Change (for Change Data Capture).

For large data-driven companies, this problem is especially acute. With sufficient resources, it is not surprising that many of them have developed their own tools, many of which are our competitors.

Around the world, almost any solution has its analogues, especially in the IT field. Therefore, for a detailed description and correct implementation of our DataTool product, it is necessary to determine the significant pros and cons of Let us highlight three key potential competitors and compare their features and qualities.

- Amundsen (Lyft)
- DataHub (LinkedIn)
- Metacat (Netflix).

In short, these solutions consist of three main blocks—Ingestion layer (receiving metadata from various sources in DWH), Metadata storage and indexing (storing metadata for the selected model, test search functionality) and RestAPI and UI (displaying results and information). Ready-made connectors are also an important point. Important is the model of the Ingestion layer block with metadata. It can work on two models—push (when the source itself reports the appearance of a new table or schema change through a message in Kafka) and pull (when the scheduler itself makes requests at certain intervals to obtain metadata from the source).

Compare characteristics (see Table F.1 of Appendix F).

- **Amundsen**
  - **Pros**: the largest list of supported sources, among which is Snowflake, the ability to use dashboards in the metadata model, ease of deployment, the main stack is python and airflow
  - **Cons**: no tracking of dataset changes (keeps only the current state and works as a reference), no data lineage (the source can only be identified in the block with custom information)
- **DataHub**
  - **Pros**: pull and push models (to receive changes in the schema in real-time), a huge number of supported connectors, Kafka topics as sources
  - **Cons**: in the public version, there is nothing in the data model except the table and users; the relatively complex structure of the tool; the complexity of the deployment; the overloaded UI
- **Metacat**
  - **Pros**: larges number of connectors, the ability to version schemas, push notification system for table and partition changes
  - **Cons**: the complexity of extending functionality due to the java and groovy stack, lineage not implemented.

Why choose microservices?

Before turning to the description of DataTool microservices and their interaction, let us look at existing application development architectures.

Microservice architecture is a software development technique that aims to break an application down into distinct key functions, and each of these functions is called a "service". These services are designed to meet specific and unique business needs, such as user management, payments, sending emails, and notifications. In addition, they are independent and modular, with their own codebase, which allows each of them to be developed and deployed without affecting the others. All of them perform a separate part of the app. That is, each individual service contains and implies its own logic, development, construction, and certain functions. These services are built on specific capabilities and are typically deployed independently. This approach greatly simplifies understanding of the application architecture, explaining how it works to new users and developers, allows for more successful and painless integration of new features, and provides much better scalability.

Distinctive features of microservice architecture components are:

- Small independent loosely coupled services
- Business or customer scenario encapsulation
- Separate codebase for each microservice
- Service deployment independence
- Interaction with each other via API.

This definition may be reminiscent of another, very similar type of architecture, Service Oriented Architecture (SOA). What makes them different? Microservice architecture is an extension of SOA. Most of the development of microservices has already been common in the SOA world, and it is being expanded so that microservices architecture is a well-established SOA architecture. Let us make a detailed comparison based on size, reusability, communication, data duplication.

- **Size**. In a microservice architecture, a service receives a single purpose and business meaning each time, while in an SOA architecture, services can perform many business functions
- **Reusability**. One of the goals of SOA is to reuse components, while in microservices we try to keep such processes to a minimum. The reason is the emergence of dependencies that reduce flexibility and fault tolerance. The smaller the connection the better, even at the cost of duplicates
- **Communication**. In SOA, communication between services is typically synchronous over an enterprise service bus (ESB), which creates a critical point of failure and latency. In an application implemented using a microservice architecture, fault tolerance is better due to the independence of each microservice. There is also asynchronous communication between services, which allows for high availability
- **Data duplication**. One of the goals of SOA is to allow applications to access data synchronously at the source, which slows down the communication and creates the possibility of queues from processes. In a microservice architecture, each service has access to all the data it needs locally to ensure maximum performance and flexibility, even if that means duplication of data and, thus, complexity.

Monolithic architecture is the opposite of microservices, since it is built as one large inseparable system—a single logical executable. With this option, as a rule, one codebase is used. Typically, an enterprise application consists a database, a client-side user interface, and a server-side application. A monolithic application is most often deployed at the same time, frontend i.e., client and backend server, regardless of what has been changed. To make any changes, the employee must develop and deploy an updated version of the server application.

Let us see what are the microservice advantages and drawbacks [17]. The cons of microservice architecture are—Complexity, Testing, Data Integrity, Network latency:

- **Complexity**. Indeed, a single service may seem simple, but the whole system will be more complex. Each of the components must be unwrapped independently
- **Testing**. The more services, the more difficult it is to write tests for the entire application. However, the microservice approach to development makes it convenient to test changes within one service, the entire application, requires thorough testing especially in use of dependencies between services
- **Data Integrity**. Microservices have a distributed architecture, much like a distributed database. Therefore, there can be problems with data integrity. Some business entities that may need updates to multiple business functions of an

application require multiple databases or partitions of the same database to be updated, and different services may be attached to them. This in turn requires data consistency tuning, which is obviously more complex and less intuitive for developers

- **Network latency**. The more services, the greater the number of interactions between them, respectively, the chains of dependencies of services to solve the set of business tasks become more and more difficult, and the time delay becomes longer. The remedy, can be asynchronous data, exchange, however, complicate the system.

There are certain cons, and often are quite significant. Nevertheless, microservice architecture is referred to as best practices. Let us describe in detail what the reasons are—independent development, independent deployment, independent scaling, small targeted teams, small code base, data isolation, resilience, and technological advances.

- **Independent development and Small Targeted Teams**. For each microservice, a different technology stack can be chosen best suited for the implementation of a business task. A single team can also be defined, i.e., only one specific group of employees can create, test and deploy the service, which speeds up deployment and promotes continuous innovation. In addition, it is more difficult to make a mistake as there are clear boundaries between different services
- **Independent deployment**. If, in applications built on a monolithic architecture, one mistake can block the entire release process then, in the case of microservices, everything is different. They are independently deployed and can be updated without redeploying the entire application, making it easier to manage bug fixes and new features, integrate, and continuously deploy
- **Independent Scaling**. At this point, the conditional independence of services also comes to the fore. By optimizing cost and time, they can scale separately from the entire application according to needs or business requirements
- **Small Code Base.** In a monolithic application, adding new features typically requires making changes in many places because code dependencies tend to mix up over time. In a microservice architecture, these dependencies are kept to a minimum, making it easy to add new features, understand code, and introduce new team members
- **Data Isolation**. In a microservice architecture, each service has private access to its database, which allows to update the database schema without affecting other services. In a monolithic application, updating the schema can become very complex and risky because multiple parts of the application may use the same data
- **Resilience**. With a microservices architecture, critical points of failure are considerably reduced. When a service goes down, the whole application does not stop functioning as it does with the monolithic model and, therefore, the risk is also reduced when new features are developed. Errors are isolated and therefore easier to correct

- **Technological Advances**. Recent advancements in cloud technologies and containers make setting up a microservices architecture increasingly simple.

In microservices certain services must interact with each other, this can be achieved in two ways: orchestration or choreography.

a. **Orchestration.** The first intuitive idea to get services to communicate with each other is to simply make REST calls between services, as shown in Fig. F.1 of Appendix F.

This approach is easy to set up, but the system quickly becomes more difficult to maintain the dependencies between the services. A smarter way to do this transaction is to introduce an additional abstraction with a new service, the orchestrator [17] (see Fig. F.2 of Appendix F).

The orchestrator "knows" about the other services. Service upgrades need care not to "destroy" the orchestrator. However, this approach still has drawbacks, such as the latency that can result from synchronous calls and poor fault tolerance. Also, a system consisting of a large number of services has many dependencies that look like a distributed monolith. Therefore, orchestration should be used with caution (see Fig. F.3 of Appendix F).

b. **Choreography**. The choreography overcomes certain shortcomings of the orchestration, and there is less dependency or delay in the system. The solution is to use events with a publish-subscribe model.

When the action is completed, the service publishes an event that other services can subscribe. Communication is asynchronous and the services are unaware of each other, so the system is efficient and easy to maintain on a large scale. In the event of a service failure, the system continues to operate, but data consistency may be lost. That is why high system availability is preferred, in cases where data consistency is critical, and it is quite possible to install circuit breakers to automatically disable certain services.

Each service shares logs and metadata with the database, and that each of them has more functions than shown. The arrows indicate the priority of actions top down. Dashed arrows show indirect user actions. That is, for example, the recommendation service independently reads the user actions.

The diagram does not show connections with the services described in [13].

The of services use the Python programming language. Many developers choose this language to implement services, due to suitable frameworks [20].

The choice of technology was FastAPI [21]. FastAPI is a framework for creating concise and rather fast HTTP API servers with built-in validation, serialization, and asynchrony. It stands on the shoulders of two other frameworks: Starlette for web, and Pydantic for validation. The UI uses Bootstrap [23] and jQuery [15].

Let us start with the fundamental microservice for the section, Data Engineering microservice.

6.1.2 Data Engineering Microservice

Currently, there is no universal solution available to connect the orchestrator or ETL tool with other apps or business processes and also work with metadata. That is, users constantly run between different applications. The microservice-based implementation follows.

Data Engineering microservice provides storage, execution, and implementation of other tasks. Its job is to bridge the existing ETL tool, the git repository, and the rest of the system, and provide actions on metadata. This, ensures the convenience of working with scripts, pipelines, and tasks. An ETL tool is Apache Airflow [8]. That allows to develop, plan, and monitor complex workflows. The main feature is the Python programming language for the processes. Airflow is used as an orchestrator for ETL/ELT processes.

The main workflow elements of Apache Airflow include:

- Directed Acyclic Graphs (DAG)
- Scheduler
- Operators
- Tasks.

According to Enlyft [19], at least 859 companies use this solution.

The Data Engineering service stores executable code and implements two REST API methods.

- `GET/repo`: returns the external link to the repository
- `POST/repo`: creates a new repository based on a template.

The task of this service is to create new git repositories with a pre-prepared structure of files, configuration objects (via `pydantic BaseSettings`), requests that return preset values (using `FastAPI Router`), and implement integration with Apache Airflow instance (simultaneously providing `gitsync` for `git repo` and `dags` scripts). To implement these points, a git server with a separate container (`dockercompose`) is deployed.

This service creates templates, git repository associated with the specific Apache Airflow instance for the data pipelines.

A git server is deployed in the docker container, which allows us to create new repositories with a single command (`docker-compose exec git-server mkrepo airflow`). Also, with the help of Kubernetes, the Apache Airflow tool is installed (Fig. 6.1), to which, using the `gitsync` function, users can synchronize the created repository. Users create templates by specifying the name of the repository.

The microservice operation is shown in Fig. 6.2. Let us recall the interaction between microservices (see also Fig. F.4 of Appendix F).

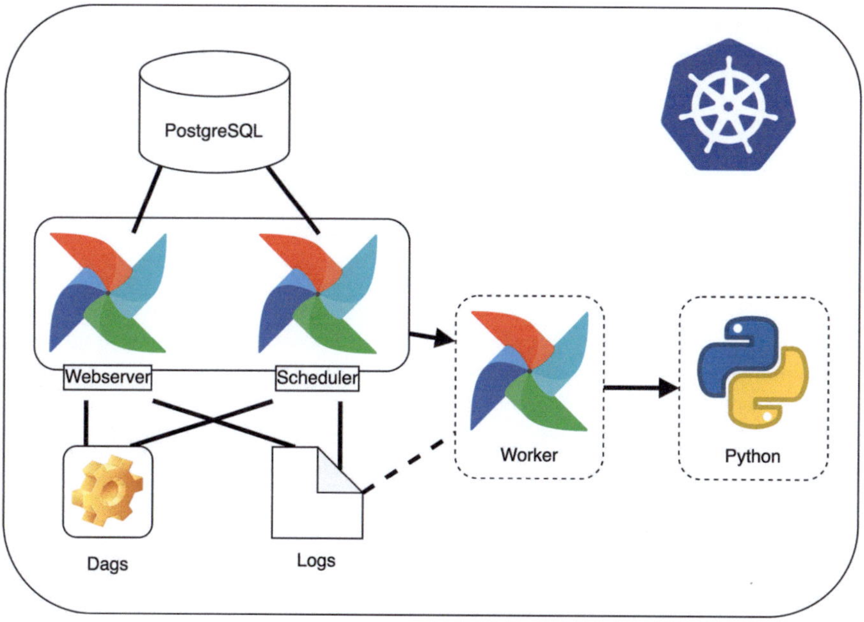

Fig. 6.1 Apache airflow on kubernetes

Fig. 6.2 Data engineering microservice

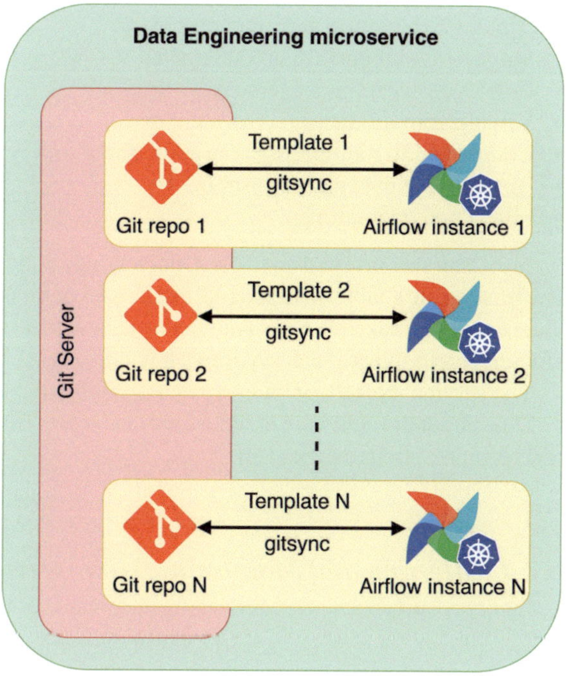

This service is a "subscriber" to the Business Intelligence service.

- Monitoring and alerting microservice to comply with the principles of data quality
- Recommendation feature service that uses the collected information about the user and other data to process and provides recommendations.

6.1.3 Business Intelligence Microservice

The task of this microservice is to contact the embedded database, create a connector, get the necessary information and let the developers play with its visual representation.

For visualization, the choice was *Redash* [18]. This service interacts with users by means of a Pub/Sub bus (see Fig. 6.3.1 of Appendix F). *Redash* helps to work with databases, and by connecting them to the service users are able to process the available data using SQL queries and convert them into graphs and charts.

Ready-made graphic data is placed on the dashboard, a panel of several graphs and textual data showing analytical information from the requested databases. Dashboards can be shared with colleagues or made public.

Redash features are:

- Queries to data sources
- Schema exploration and autocompletion
- Snippets for frequently used elements
- Employee access management
- Visualization of data
- Connecting to multiple sources at the same time
- Auto-update data on the dashboard
- Filters and data settings.

Metadata are inserted into the Query window by scripts that allow displaying information in a table without additional user actions. In this case, employees can independently make changes and gather information from the database much more efficiently and quickly than by Redash itself.

The service interacts with users by Pub/Sub bus, (see Fig. F.6 of Appendix F).

Data about the activities of the BI service is processed by the monitoring, alerting and recommending mechanism.

6.1.4 Alerting and Monitoring Microservice

Many software developers spend a lot of time implementing pipelines and similar function. However, the team often, stops there and does not monitor and alert.

Although, effective monitoring and alerting helps to enhance efficiency of the data-driven workflows.

The main task of monitoring and alerting is for the team to notice implementation problems before they get hit on the head by a business user. In Big data, such processes help to notify employees about problems in the implementation of pipelines, mapping fields data reporting, and so on. Many tools already have built-in alerting functionality, such as the *Fivetran* cloud ETL solution (see Fig. F.7 of Appendix F) and the *Redash* visualization tool (see F.8 of Appendix F).

In alerting and monitoring, the central concept is Data Quality (DQ) defined as the data suitability for a given purpose affecting.

- Trust in the data used
- Quality of management decisions
- Customer satisfaction
- Effective marketing
- Cost reduction/margin increase
- Regulatory and reputational risks.

Further, requirements form the basis of the Alerting and Monitoring microservice according to the following DQ evaluation criteria.

- **Completeness**: data gaps. The service must compare input and output data, to ensure absence of leaks
- **Uniqueness**: The service must monitor the quality of the information that the user operates and avoid duplicates in the metadata, which can eventually grow into errors. This can also affect the performance of microservices, most of which depend on the information provided by neighbors
- **Referential integrity**: Especially important in the DE service because, due to errors, the users may lose data
- **Validity**: Compliance with the rules. Verification of compliance with the parameters of the operation of services in terms of consumed resources, business requirements and principles embedded in the implementation paradigm
- **Timeliness**: Relevance to a moment in time. This means versioning and ordering relationships of the schema elements (see also Fig. 6.2).

The task of this microservice is to set up alerts and monitoring on the functions implemented including DE, BI, Recommendation function, Data architecture and Database sync microservices; with the components that:

- Adjust peak values for various parameters
- Visualize and see similarities or differences between data before and after the pipeline execution (check data completeness)
- Receive notifications of faults or warnings [13].

Grafana [3] was chosen as a tool for data visualization, the connector to be implemented in this service. Grafana is an open-source data visualization, monitoring and analysis platform. It allows users to create dashboards with panels, each displaying specific metrics over a period of time. Each dashboard can be customized for specific development and/or business needs.

The service includes PUB/SUB bus and to store monitoring parameters (see Fig. F.9 of Appendix F).

This service receives data from all other services in the DataTool, it is a "dependent subscriber".

6.1.5 Chats: Text Messaging and Notifications Microservice

The idea for using this microservice in the solution from JetBrains Space [16]. Space is a universal solution for a software development team. It represents the entire development lifecycle, from hosting Git repositories and securing the CI/CD pipeline to publishing packages. However, the part responsible for communication between colleagues, in a chat, is of the greatest interest (see Fig. 6.3 of Appendix F) [16].

This microservice (see Figs. F.11 and F.12 of Appendix F).

- Communicates on tasks without interrupting the implementation, access to tools, pipelines and data
- Searches among application users within the team or company, depending on the desire of the client
- Exchanges of text and media files (pictures, videos, documents, etc.)
- Receives notifications about an event, a message.

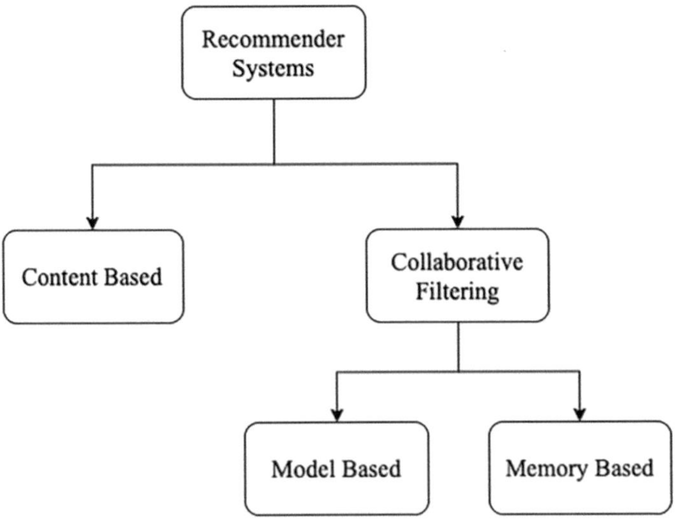

Fig. 6.3 Types of ML recommender systems

After registering the user, text messages can be placed in the dialog box.

Functionally, the service backend implements:

- Socket manager, i.e., class responsible for sending messages between the connected users and maintaining their connections
- Web-socket endpoint to call the manager
- Register and user endpoint.

The frontend and the backend were integrated by templates from the FastAPI app.

Concerning interaction of the microservices, the users exchange messages, and the service reflects events in Pub/Sub and records the history.

This service does not interact directly with other services as Pub/Sub bus and recommendation microservice (see Fig. F.13 of Appendix F) [1].

6.1.6 Recommendation Function Microservice

The recommending systems make quite a large class of ML algorithms that offer users certain functions or further actions. Large recommender systems are widespread in retail, video, music streaming, etc. (types are shown in Fig. 6.3).

A recommendation system for software engineering (RSSE) is an application that provides information items estimated to be valuable in a given context. RSSE typically visualize source code artifacts. They aim to improve system quality, make the development process more efficient and less expensive, lower cognitive load, and help to make better decisions. They mainly support reuse and debugging, implementation, and maintenance [14].

The task of this microservice is to provide intelligent tips for application consumers using ML methods.

The data is collected and stored in a separate database structure for further training of the ML model and application in DataTool.

Data preprocessing. In this phase, techniques and tools extract valuable information from different data sources. In particular, structured data adheres to several rules that organize elements in a well-defined manner. Source code and XML documents are examples of this category. Unstructured data may represent different content without defining a methodology to access the data. Documentation, blog, and plain text fall into this category. Thus, the data preprocessing component must be carefully chosen to fit the features of these miscellaneous sources.

Capturing context. After the data preprocessing phase, the developer context is excerpted from the environment to recommendation engine. A well-founded technique primarily employed in the ML domain is the FeatureExtraction, to concisely represent the developer context. Principal Component Analysis (PCA) and Latent Semantic Analysis (LDA) are just two of such techniques. Keyword extraction and APICallExtraction are two techniques mostly used when the Capturing Context phase has to analyze source code. Capturing context often involves a search over big software projects. Thus, a way to store and access a large amount of data is needed to speed up the recommending. Indexing is a technique mainly used by the code search engines to retrieve relevant elements.

Producing recommendations. In this phase, recommendation algorithms produce suggestions relevant for the context. Concerning data mining techniques, some of them are based on pattern detection algorithms. Clustering is usually applied to group objects according to some similarity functions. The most common algorithm is the K-means based on minimizing the distance among the items.

Presenting recommendations. Uses severed strategies extensions for IDEs and dedicated Web-based interfaces.

Presenting recommendation uses such strategies as extensions for IDEs and specific web interface (see Fig. F.14 of Appendix F).

6.2 Conclusion

In this chapter, we conducted a case study to develop a set of data management microservices. The future plans are to integrate these with the microservices of the DataTool product, and to create a mobile version for this tool.

References

1. Zheleznoy, A.: Smart database orchestrator for microservice-based applications, term project, p. 2021. Comp. Sci., HSE Univ., Moscow, Russia, Fac (2021)
2. Dyakov, R.: Software architecture for express data integration in enterprise corporate systems, term project, p. 2022. Comp. Sci., HSE Univ., Moscow, Russia, Fac (2021)
3. McCollam, R.: Getting started with grafana: Real-time dashboards for monitoring business Operations, 1st ed. Apress, (2022)
4. di Rocco, J., di Ruscio, D., di Sipio, C., Nguyen, P.T., Rubei, R.: Development of recommendation systems for software engineering: the CROSSMINER experience. Emp. Soft. Eng, **26**(4), (2021)d. https://doi.org/10.1007/s10664-021-09963-7
5. Clements, P., Gordon, P., Kazman, R., Klein, M.: Evaluating software architectures: Methods and case studies, (1st ed. Addison-Wesley Professional, (2021)
6. Runner SaaS|GitLab.: GitLab. (2022). https://docs.gitlab.com/ee/ci/runners/. Accessed on 10.05.2025
7. Malcher, M., Curtis, B., Lawless, C.: Oracle data integration: Tools for harnessing data, 1st ed. McGraw Hill, (2015)
8. Harenslak, B.P., Ruiter, D.J.R.: Data pipelines with apache airflow. Manning, (2021)
9. Airbyte.: Open-source data integration platform. ELT tool. (2022). Airbyte. https://airbyte.com/. Accessed on 10.05.2025
10. Kuhn, D., Kyte, T.: Expert oracle database architecture: Techniques and solutions for high performance and productivity, 4th ed. Apress, (2021)
11. Riggs, S., Ciolli, G.: PostgreSQL 14 Administration Cookbook: Over 175 proven recipes for database administrators to manage enterprise databases effectively. Packt Publishing, (2022)

12. R, V.A.: Up and running with clickhouse: Learn and explore ClickHouse, Its robust table engines for analytical tasks, ClickHouse SQL, Integration with the ClickHouse Server. BPB Publications, (2021)
13. Dyakov, R.: Microservices for Warehouse and Business Intelligence Management in Data Integration, master's dissertation. Comp. Sci., HSE Univ., Moscow, Russia, Fac (2022)
14. Gasparic, M., Janes, A.: What recommendation systems for software engineering recommend: A systematic literature review. J. Syst. Softw. **113**, 101–113 (2016). https://doi.org/10.1016/j.jss.2015.11.036
15. Nixon, R.: Learning PHP, MySQL & JavaScript: With jQuery, CSS & HTML5 (Learning PHP, MYSQL, Javascript, CSS & HTML5), 5th ed. O'Reilly Media, (2018)
16. Chats: Text Messaging and Notifications. JetBrains Space.: JetBrains Space Help. (2021). https://www.jetbrains.com/help/space/chats.html. Accessed on 10.05.2025
17. Bouttier, S.: Microservices architecture from A to Z—The startup. Medium. https://medium.com/swlh/microservices-architecture-from-a-to-z-7287da1c5d28. Accessed on 10.05.2025
18. Setting up An Alert. Redash Help. (2021). https://redash.io/help/user-guide/alerts/setting-up-an-alert/. Accessed on 10.05.2025
19. Apache Airflow commands 0.45% market share in Data Integration.: Companies using apache airflow. https://enlyft.com/tech/products/apache-airflow
20. Five Best Technologies to build microservices architecture. Clariontech (2021) https://www.clariontech.com/blog/5-best-technologies-to-build-microservices-architecture. Accessed on 10.05.2025
21. Feofanov, I.: Getting to know FastAPI Habr. (2020). https://habr.com/en/post/488468/. Accessed on 10.05.2025
22. Daeppen, M.: Data governance beyond SDX: Adding third party assets to Apache Atlas. Cloudera Blog. (2021). https://blog.cloudera.com/data-governance-beyond-sdx-adding-third-party-assets-to-apache-atlas/. Accessed on 10.05.2025
23. Lett, J.: Bootstrap 4 quick start: A Beginner's guide to building responsive layouts with bootstrap 4 (Bootstrap 4 Tutorial) (Illustrated ed.). Bootstrap Creative. Accessed on 10.05.2025

Chapter 7
The Digital Lessons: Patterns and "Antipasti"

Abstract This final chapter summarizes the outcomes of the book using the methodology and framework of IT Crisisology. It starts by discussing the previous chapter findings which incorporate IT intensive startups, complex digital production, versatile businesses and other digitalization cases. The focus is combining agile practices and cutting-edge techniques required to manage the complexity of digital production crises and promote sustainability. Although the individual solutions may significantly differ depending on project scale and scope, this chapter demonstrates that common patterns may be obtained and systematically applied to promote sustainability. In addition, the chapter discusses the dramatic anti-patterns resulting from disproportionate rigidity and confusing human factors as well as a few other potential crisis triggers. Overall, the chapter paves way to systematic digital success strategies and crisis-proof, sustainable product design for the multiple contexts being addressed.

7.1 Lifecycle Plaining in Changing Landscape

The methodology for planning the lifecycle of information systems, in the context of a rapidly changing technological landscape, requires flexibility, adaptability and efficiency in decision-making. The goal of the methodology is to ensure the development, deployment and management of information systems that can quickly respond to new technologies and business requirements.

Each stage contains the key processes and the tools, methods, and technique uses [1–7].

These points structure the description of the methodology, providing general understanding and specific details for successful implementation.

S. V. Zykov, *IT Crisisology Patterns and Practices*, Smart Innovation, Systems and Technologies 449, https://doi.org/10.1007/978-981-95-1260-7_7

7.2 Stages and Recommendations

Let us summarize stages of the methodology for planning the lifecycle of information systems for a changing technological landscape (see Fig. 7.1).

Stage 1. Analysis of requirements and assessment of changes in the technological landscape.

Objective: to identify the key requirements for the system and to take into account the impact of current and future technologies on system planning.

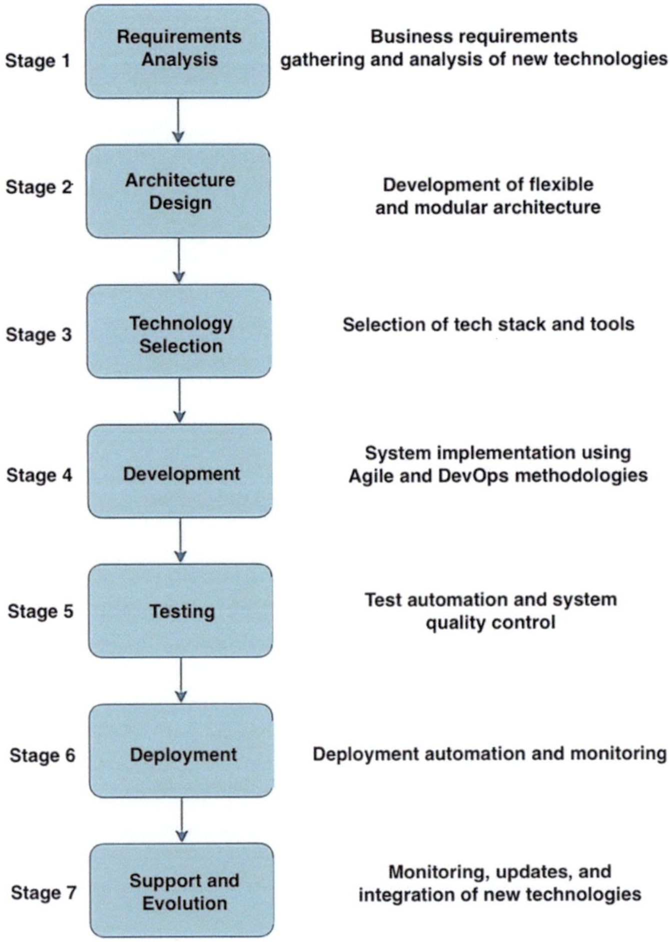

Fig. 7.1 Methodology stages

Description: At this stage, requirements are collected from all stakeholders (business, end-users, and IT department), etc. At the same time, the current technological landscape is analyzed, including technologies such as cloud computing, artificial intelligence, blockchain, and automation. Operational analysis of requirements allows to predict changes several years in advance and design the system ready to adapt and scale to new technologies such as AI or IoT.

Key processes

- Definition of business goals and needs
- Analysis of the possibilities and limitations of current technologies
- Forecasting future changes in technology
- SWOT analysis to assess the impact of new technologies.

Tools

- Jira for requirements management
- Confluence for documentation
- SWOT analysis allows to take into account the strengths and weaknesses of the system, identify possible threats and opportunities associated with the introduction of new technologies
- Google Trends or Gartner Hype Cycle help track and predict technology trends, respond to future potential changes timely.

Methods

- Agile analysis and user stories help in the iterative managing requirements, especially in a changing landscape. This allows you to quickly adapt the system to new user needs
- PEST (Political, Economic, Social, and Technological Factors) analysis helps to assess the impact of external changes on business.

Comments: For existing systems, review the current architecture and technologies to determine whether they are ready to integrate new solutions. Upgrades to individual components or architectural revisions may be required.

Recommendations: Collect requirements from all stakeholders, including business, IT, and end-users, to embrace critical aspects of the system. Evaluate new and promising technologies. Monitor analyze the technology market to take into account the potential for the implementation of AI, IoT, cloud solutions, automation, and other technologies. Develop change scenarios over a period of 3–5 years that may affect the system architecture and capabilities.

Stage 2. Architecture design

Objective: To create an adaptive and scalable system architecture that is ready for the integration of new technologies.

Description: At this stage, a modular architecture is developed that can easily adapt to changes. This can be either a microservice or a service-oriented architecture (SOA) that supports the dynamic addition of new modules or components. Scalability, reliability, and ease of integration of new technologies are critical. The modular architecture allows the system to be dynamically adaptive to changing requirements. For example, adding new modules to process data from IoT devices or integrating machine learning will become easier.

Key processes

- Development of modular architecture
- Definition of interaction protocols between components (e.g., API, REST, GraphQL)
- Assess the fault tolerance and scalability of the architecture.

Tools

- ArchiMate for architectural design and visualization, to identify potential bottlenecks and ensure agility
- GraphQL or REST API to facilitate interoperability between modules, keeping the system flexible and scalable.

Methods

- Microservice architecture. Enables independent module upgrades and easy integration of new technologies. A modular approach allows quick adaptive to changes;
- SOA (Service-Oriented Architecture) supports flexible integration of external services, which is important for scalability.

Comments: For older systems, refactor existing modules and move to a modular architecture with minimal dependencies between components. This will make it easier to upgrade and maintain.

Recommendations: design the system so that individual modules develop independently. This will allow for faster integration of new technologies and the modernization of the system. Develop an architecture to address the potential growth of the load and scale.

Stage 3. Choice of technologies and tools

Objective: To identify appropriate technologies and tools to build and maintain the system.

Description: the technologies that best meet the requirements of the business are evaluated, taking into account their current capabilities and future prospects. This includes choosing the platform to deploy, programming languages, databases, and external systems. Use flexible and scalable technologies, that allow to adapt to rapidly changing market requirements and implement new solutions.

Key processes

- Comparison of available technology stacks (e.g., cloud vs on-premise infrastructure, SQL vs NoSQL)
- Assessment of future technology interoperability

- Estimating the costs of implementing and supporting selected technologies.

Tools

- Docker and Kubernetes. These tools provide flexibility and ease of infrastructure management, important for dynamic systems. They support containerization and orchestration of microservices, making it easy to scale
- Terraform to manage infrastructure as code, making it easy to deploy and upgrade the system in the future
- Jenkins for CI/CD
- GitLab for code management.

Methods

- Proof of Concept (PoC). Testing technologies before implementation helps to minimize risks and make sure that the selected technology is really suitable for a particular project;

Techniques: Conducting pilot projects using PoC to evaluate the performance of the technology in real-world conditions.

Comments: Incorporating new technologies into existing systems may require hybrid solutions (e.g., using containers for individual modules or moving to a cloud infrastructure for new components).

Recommendations: when choosing technologies, focus not only on current requirements, but also on their ability to develop and maintain over a long time. Tools like Docker and Kubernetes provide flexibility and ease of infrastructure management important in a changing landscape. Test technologies by PoC and before introducing new tools or technologies, check their compatibility and effectiveness on pilot projects.

Stage 4. Development and implementation

Objective: To create a system that meets the requirements and is ready to integrate with future technologies.

Description: In this stage, the system is developed using agile methods such as Agile and Scrum. To do this, a CI/CD pipeline is created to automate development and deployment. The emphasis is on writing uniform, testable, and maintainable code. Development agility allows product changes to be made faster. Deployment automation accelerates the integration of new features, important in a changing technology landscape.

Key processes

- Building a CI/CD pipeline
- Application test and build automation
- Implement DevOps practices to ensure stability and speed up release.

Tools used

CI/CD pipeline using

- Jenkins
- GitLab CI
- Docker for containerization and automation.

These tools automate development and deployment, minimize errors and accelerate release of new features.

Methods

- Scrum, regular sprints and iterations allow to quickly respond to changes and adapt development to new requirements;
- TDD (Test-Driven Development): Helps ensure code quality and reduce the cost of fixing bugs at later stages.

Techniques: Unit testing, integration testing, automated security checks (DevSecOps).

Notes: For existing systems, adding CI/CD can significantly accelerate new feature deployment deployed and ensure that new solutions are ready to be integrated.

Recommendation: Use Agile or Scrum to ensure flexibility. This will allow to respond faster to changes and improve the product at every stage. Automate the integration and deployment process so that new features and updates can be delivered to the system faster and more securely. Align development and operations to accelerate system deployment and management.

Stage 5. Testing & Quality Assurance

Objective: To test the operation of the system in various conditions to ensure its stability and security.

Description: In this stage the system undergoes multiple testing: from unit to system and load. It is important to automate testing, to reduce manual review time and increase test coverage. Test automation significantly speeds up releasing new versions and reduces the errors. This is especially important in environments with new features added regularly.

Key processes

- Unit, integration, and system testing
- Load testing and fault-tolerance testing
- Security/IndexTerm> and compliance.

Tools

- Automated testing (Selenium for functional testing, JMeter for load testing) speeds up verification and provides high test coverage
- OWASP ZAP for system security testing and integration of secure practices into CI/CD.

Methods

- Risk-Based Testing (RBT) focuses on the most critical areas and ensures more efficient use of resources.

Techniques: automated testing, regression testing, security testing with CI/CD integration.

Notes: For existing systems, adding automation to testing detects bugs early in the integration of new components.

Recommendation: Use automated tests (unit, integration, regression) to ensure system quality and quickly validate changes. Conduct regular stress tests to assess increased load, which is important for scalable and dynamic systems. Use DevSecOps ensure that security is implemented in the lifecycle.

Stage 6. Deployment and Operations

Objective: to deploy the system and ensure its stable operation with the ability to scale.

Description: The deployment of the system should be automated to minimize the risk of errors and speed up the process. It is important to use orchestration and monitoring to manage workloads and scalability. Use deployment orchestration and automation to minimize errors, speed the update process, and ensure system stability under a changing load.

Key processes

- Automation of deployment
- Real-time system monitoring
- Incident management and automatic resolution.

Tools

- Kubernetes. Automates container deployment, management, and system scalability
- Prometheus and ELK Stack for real-time system performance monitoring and log analysis.

Methods

- DevOps
- Infrastructure as Code
- Blue-Green Deployment, to minimize downtime and risk when a new version is released.

Techniques: automated deployment and monitoring, Canary revisions to test new versions.

Notes: For existing systems, implementing monitoring and orchestration tools will help systems remain agile and stable as new components are added.

Tools/Best practice: Use orchestration and automation to deploy the system quickly and securely. Use monitoring tools to monitor system performance, respond quickly to failures or peak loads. Use flexible deployment strategies such as Blue-Green Deployment or Canary Deployment to securely deploy updates with minimal risk.

Stage 7. Support and development

Objective: To provide long-term support for the system and its adaptation to new technologies.

Description: The system must be continuously maintained, which includes regular updates and performance monitoring. The safety and optimization of operations must be monitored. Constant monitoring and automation of updates allow to respond to changes in load and infrastructure, improving the efficiency and readiness for challenges.

Key processes

- Continuous performance monitoring
- Periodic updates and security patches
- Evaluation of new technologies and their integration.

Tools used

- GitLab for release management
- Ansible to automate updates and manage configurations
- Nagios to monitor performance and detect problems timely manner.

Methods

- DevSecOps to integrate security at all levels
- Continuous Improvement (CI): continuous analysis of system performance and improvement of development, testing and operation processes.

Techniques: Incremental updates, performance monitoring, log analysis.

Comments: For existing systems, automated updates and monitoring allow to easily integrate new technologies and maintain security.

Today/Best practices: Ensure that components are regularly updated, security and new technologies adopted. Regularly review managed performance improve development, testing, and deployment processes. Continuously explore opportunities to implement new solutions that can improve system performance and efficiency.

This methodology for planning the lifecycle of information systems provides patterns and practices to adapt information system to constant changes in the technological landscape. The use of flexible development methods, automation tools, and regular monitoring makes it possible to quickly adapt and minimize risks.

7.3 Application of the Pattern

Since the full lifecycle of the IS takes very long, the demonstration of the methodology will be based on a hypothetical case.

Suppose a company is developing a new logistics management system for a large retailer. The system should integrate with external supplier services, have a modular architecture, and be ready to scale as the company plans to expand into new markets. The technology landscape is volatile, and new technologies, such as IoT for tracking goods and machine learning for demand forecasting, need to be considered.

Stage 1. Analysis of requirements and assessment of changes in the technological landscape

The company collects requirements from key stakeholders: logistics, IT, management, and analysts.

The possibility of using new technologies (IoT, AI) to automate processes and improve the forecasting of logistics tasks is analyzed. (see Tables G.1 and G.2 of Appendix G).

Stage 2. Architecture design

A microservice architecture is chosen to ensure flexibility and scalability of individual components of the system including modules for inventory management, transport, and warehouse.

The architecture supports integration with external APIs of vendors and customers.

Stage 3. Choice of technologies and tools

For development and operation, a technology stack based on Docker and Kubernetes is selected for container management.

Databases: PostgreSQL for transactional data and MongoDB for unstructured data.

Selection of integration technologies (API Gateway, MQ) for integration of microservices and external systems (see Table G.3 of Appendix G).

Stage 4. Development and implementation

Agile/Scrum methodology ensures a quick response to changes (see Table G.4 of Appendix G).

CI/CD is implemented to automate the build and deployment of applications (see Fig. 7.2).

Stage 5. Testing & Quality Assurance

Automated testing is used at all levels.

Particular attention is paid to performance and fault tolerance testing. (see Table G.5 in Appendix G).

Stage 6. Deployment and Operations

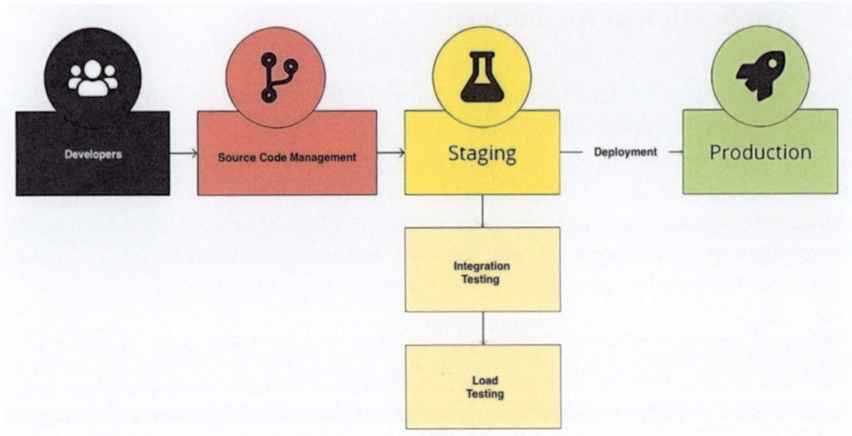

Fig. 7.2 Diagram of the CI/CD process

Deployment follows a Blue-Green strategy to avoid downtime and minimize the risk of upgrades. Blue-Green Deployment contains two identical environments: Blue and Green. The current working environment is Blue. The new version is deployed to the Green environment, is tested and, if successful, switches to it.

Real-time monitoring of the system using Prometheus and Grafana.

Stage 7. Support and development

Regular system updates, security improvements, and the addition of new features based on user feedback (see Table 7.1).

Continuous performance improvement through resource monitoring and optimization.

Monitoring panel (Grafana)

- Metrics: CPU usage, request latency, number of active users
- Setting alarms when critical thresholds are reached

Difference from existing models

1. Focus on the Changing Technology Landscape

The proposed methodology pays special attention to the crises rapid change of technologies (IoT, AI, DevOps, cloud solutions) and their impact on the architecture and

Table 7.1 Update schedule

Renewal	Frequency	Comments
Security patches	Monthly	All security updates are released on time
Code optimization	According to the results	Monitoring-based system optimization

processes of IS development. Unlike Waterfall, the proposed methodology constantly evaluates and integrates new technological solutions.

2. Modular architecture and microservices

Important features are design of a system with a modular architecture and the use of microservices. This allows the system to quickly adapt and scale as requirements and technologies change, which is not supported in traditional linear models such as Waterfall.

3. Automation and CI/CD

In the proposed methodology, the key aspect is the implementation of automation and continuous integration (CI/CD). Unlike traditional practices, where automation can be minimal, this technique involves automation at all stages including development, testing and deployment which is especially important in rapid technology change.

4. Flexibility and adaptability

Unlike strictly linear Waterfall, the proposed methodology includes agile approaches such as Agile and DevOps, allowing to quickly adapt to changing conditions. It aims at an iterative process with the ability to make changes at any point in the lifecycle.

5. Integrate new technologies at every stage

The methodology specifically includes the stages of analysis of new technologies and their integration, which is not typical for traditional models. It allows to continuously evaluate and implement cutting-edge solutions such as cloud computing and machine learning into existing infrastructure.

6. Support and continuous development

In the proposed methodology, much attention is paid support and continuous improvement, which allows the integration of new technologies. This significantly distinguishes the methodology from traditional approaches, where support is often static.

7.4 Conclusion

This chapter presented a methodology for the lifecycle of IS, taking into account technological changes. The analysis of existing lifecycle models and development methods showed that many of them are not flexible enough and are not always adapted to the conditions of a rapidly changing technological environment, which can lead to the obsolescence of systems.

The trends, such as the widespread use of cloud technologies, artificial intelligence, and big data confirmed the need to develop an approach that can quickly respond to changes. Key factors affecting IS efficiency were identified, such as new security standards, scalability requirements, and the need for continuous integration of innovations into systems.

The proposed IS lifecycle planning methodology offers a modular and flexible approach that allows new technologies to be introduced as they become available and keeps systems relevant in the long term. The methodology highlights the stages of requirements analysis, architecture design, technology selection, development, testing, deployment, and ongoing support to minimize the risk of obsolescence and improve system performance.

Based on the proposed methodology, recommendations have been developed to help organizations adapt their information systems to changing conditions. These recommendations are designed to increase agility, accelerate development and deployment processes, and ensure that the systems are resilient to change.

The proposed methodology and recommendations will help adapting IS to the dynamically changing technological landscape, which is important in competitive environment.

References

1. Laudon, K.C., Laudon, J.P.: Management information systems: Managing the digital firm. Pearson, (2016)
2. Miller, J., Davis, L.: Integrating DevOps into SAFe environments, (2023)
3. Brodie, M.L., Dang, Q., Liu, L.: On the evolution of information systems. Commun. ACM **60**(8), 48–54 (2017)
4. Petersen, K., Wohlin, C., Baca, D.: The waterfall model in large-scale development. Eur. J. Inf. Syst. **18**(4), 372–383 (2009)
5. Zhu, Q., Meng, W., Yin, J.: Integrating user feedback into iterative software development: A new approach. Int. J. Inf. Manage. **56**(2), 102–117 (2021)
6. Jiang, H., Li, Y., He, K.: Enhancing the spiral model with a fuzzy logic-based risk management approach. IEEE Access **7**(2), 15003–15016 (2019)
7. Highsmith, J.: Agile project management: Creating innovative products. Addison-Wesley, (2019)

Conclusion: Is There a Life After Digitalization?

The final chapter presented a systematic set of handy patterns and useful practices to conquer the digitalization crisis. Now, let us digest the summary of the book outcomes and revisit the introductory story of the Eiffel Tower. On the one hand, there is a great idea of a tremendous and very complex object. On the other hand, there is a set of massive techno-driven disasters similar to the catastrophe at the Chernobyl Nuclear Power Plant and Kursk submarine tragedy as possible transformation results. How to avoid these failures? (Fig. 1).

Fig. 1 Eiffel tower: **a** General view; **b** Construction details. (*Source* Zykov S.V., 2025)

S. V. Zykov, *IT Crisisology Patterns and Practices*, Smart Innovation, Systems and Technologies 449, https://doi.org/10.1007/978-981-95-1260-7

Typically, any system development lifecycle is a complex and multi-level process that requires a set of interrelated activities. In our case, the IT Crisisology framework guides this process by means of models, methods, and tools. This process usually starts from sketching and building a high-level model of the future system. Instances of such models are given in Chap. 1 and a dedicated book [1]. Such modeling was perhaps the initial step that Eiffel used when building the process for the tower in Paris. However, even at this early stage, the French genius obviously reused the parts of his previous bridge models; otherwise, the process would take longer and result in much lower quality deliverables. To implement this tower's complex project consisting of millions of individual parts, we would recommend the "divide-and-conquer" approach.

The stage that follows modeling is design which, with this approach in mind, starts with a high-level perspective and progresses to the detailed layout. In both traditional construction and software development, this stage is called architecting. According to the ITC framework, system architecture should meet requirements for three main groups of the constraining business, technical, and human-related factors [2]. For example, the technical factor would include shapes and sizes for the rivets and other tiny elements of the tower. An instance of the business factor would be shaping the monument to commemorate the ideas of Liberty and the French Revolution. The human part would include the arcs at the bottom of the tower which seem a pretty entrance decoration for the World Exhibition, although technically redundant. Chapter 2 specifically addresses the architectural issues such as migrating to microservices which in a way resemble smaller and multi-level components for the tower as opposed to monolith-like St. Peter's Cathedral. Again, pattern-based strategic reuse was a lesson to learn at the design stage as well.

Construction stage included making the elements and assembling them into bigger units to be integrated into the 300 m-high tower. This again requires a thoroughly planned process with the deliverables and milestones as well as the tools to support the useful patterns and best practices previously applied elsewhere. The further Chapters describe these processes adjusted for certain subject areas such as oil-and-gas production, portfolio project management, IT servicing, and data integration.

The concluding part of the lifecycle is tailoring the implementation to customer intimate needs as addressed in the following Chapters. Same as above, such best agile practices as Scrum, DevOps etc. discussed in the final Chapter and the related case studies appear to help and finalize the crisis responsive toolkit [3].

Summarizing the takeaways and recalling the amazingly cost-effective Eiffel Tower project, let us suggest active application of the helpful patterns and handy practices systematically presented in this book. This volume being a "Swiss army knife", as the authors sincerely hope, will assist in efficient and crisis-resilient construction of the new digital world and development of the next-generation post-digital products..

References

1. Zykov, S.V.: IT crisisology models: Object-based optimization for sustainable development., vol. 381: Springer Series in Smart Innovation, Systems and Technologies, p. 195. Springer, Switzerland (2024). ISBN 978–981–99–8648–4
2. Zykov S.V.: IT crisisology: Smart crisis management in software engineering. models, methods, patterns, practices, case studies issue 210. Springer, Singapore (2021). ISBN 978–981–33–4434–1, https://doi.org/10.1007/978-981-33-4435-8
3. Zykov S.V.: IT crisisology casebook: Smart digitalization for sustainable development, vol. 300. Springer, Singapore (2022). ISBN 978-981-33-4434-1. https://doi.org/10.1007/978-981-19-2231-2

Appendix A
Crisis Trigger or Remedy?

See Table A.1.

Table A.1 Comparison of software development methodologies

Methodology	Efficiency	Flexibility	Alignment with business strategies
Agile	Highly effective in dynamic projects with constant feedback	Extremely flexible, allowing for continuous change	Strict adherence to customer-centric innovation strategies
Scrum	Effective for complex projects in a short time	Flexibility during a sprint, but to a lesser extent in the middle of a sprint	Fits strategies that require frequent small releases and feedback
Lean	Effective for optimizing processes and eliminating waste	Moderately flexible, efficiency-oriented	Best suited for companies looking to improve operational efficiency and maximize profits
Kanban	Effective for continuous work management	High flexibility ensures a continuous flow	Ideal for businesses focused on continuous supply and efficiency
DevOps	Highly effective for fast, reliable release and system stability	High flexibility thanks to automation and CI/CD	Aligns with strategies that prioritize speed, reliability, and innovation
SAFe	Effective for scaling Agile in large enterprises	Moderate flexibility, balanced structure with flexible practices	Aligns with corporate strategies that require coordination between multiple teams and large projects
Waterfall	Effective for projects with stable and clear requirements	Minimal flexibility, rigid sequential structure	Suitable for businesses that require predictability and clear time frames

S. V. Zykov, *IT Crisisology Patterns and Practices*, Smart Innovation, Systems
and Technologies 449, https://doi.org/10.1007/978-981-95-1260-7

Appendix B
Architecting Transaction Management: From Monolith to Microservices

See Tables B.1, B.2, B.3, B.4 and B.5 and Figs. B.1, B.2, B.3, B.4, B.5, B.6, B.7, B.8, B.9, B.10, B.11, B.12, B.13, B.14 and B.15.

Table B.1 Comparison between traditional SOA and microservices

	SOA	Microservices
Inter-service communication	Smart pipes, such as Enterprise Service Bus, using heavyweight protocols, such as SOAP and the other web services standards	Dumb pipes, such as a message broker, or direct service-to-service communication, using lightweight protocols such as REST or gRPC
Data	Global data model and shared databases	Data model and database per service
Typical service	Larger monolithic application	Smaller service

Table B.2 Microservice quality attributes based on literature review

	Factor	[7]	[8]	[11]	[13]	[24]	[25]
1	Coupling	+	+	+	+	+	+
2	Cohesion	+	–	+	+	+	+
3	Autonomy	+	–	–	–	–	–
4	Granularity	+	+	+	+	+	+
5	Reusability	+	–	–	+	–	+
6	Abstraction	+	–	–	–	–	–
7	Complexity	-	–	–	+	–	–

S. V. Zykov, *IT Crisisology Patterns and Practices*, Smart Innovation, Systems and Technologies 449, https://doi.org/10.1007/978-981-95-1260-7

Table B.3 Benefits and drawbacks of REST technology

Benefits	Drawbacks
It is simple and commonly known	It only supports the request/response communication style
High testability of HTTP-based API in a browser (e.g., by Postman plugin), or in a command line (e.g., by curl, assuming JSON or other text format used)	Reduced availability. Due to direct client-to-service communication without any media to buffer messages, they must both run for the entire exchange time
Direct support of request/response style communication	Clients must know the locations (URLs) of the service instances(s) and must use service discovery mechanism to locate service instances
HTTP is firewall friendly	Fetching multiple resources in a single request is challenging
Intermediate broker is optional, which simplifies system architecture	Mapping multiple update operations to HTTP is often difficult

Table B.4 Benefits and drawbacks of gRPC technology

Benefits	Drawbacks
Designing API with a rich set of update operations is straightforward	JavaScript clients need more labor to integrate with a gRPC-based API than with REST/JSON-based one
Efficient and compact IPC mechanism, especially for large messages	Older firewalls might not support HTTP/2
Bidirectional streaming enables both RPI and messaging styles of communication	
Promotes interoperability between clients and services for a wide range of languages	

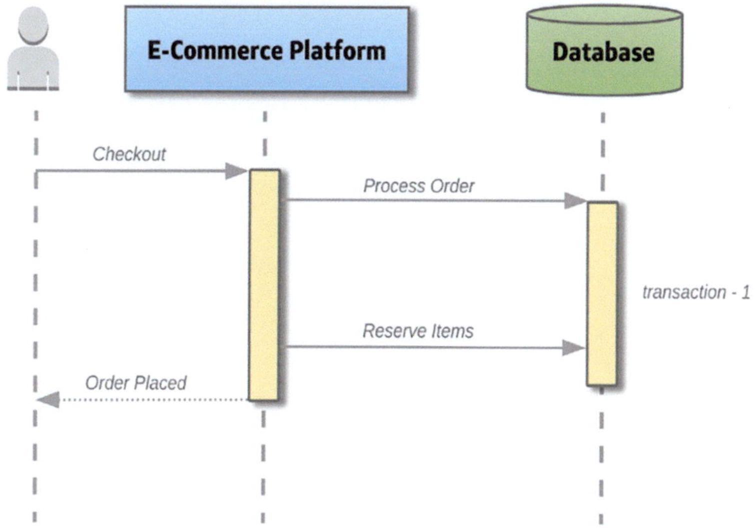

Fig. B.1 Monolith transaction sequence

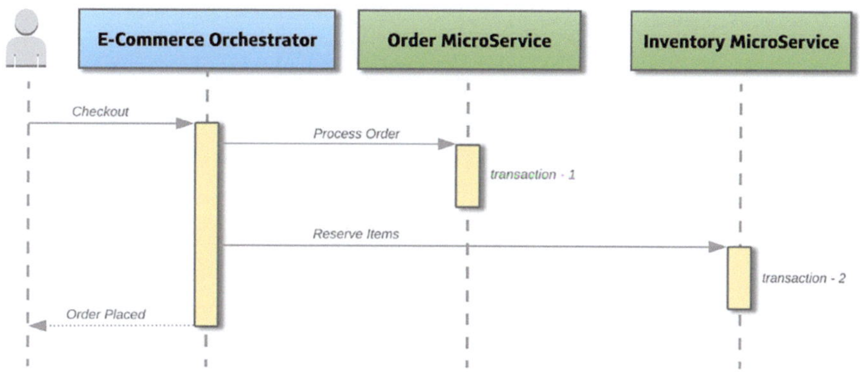

Fig. B.2 Microservice transaction sequence

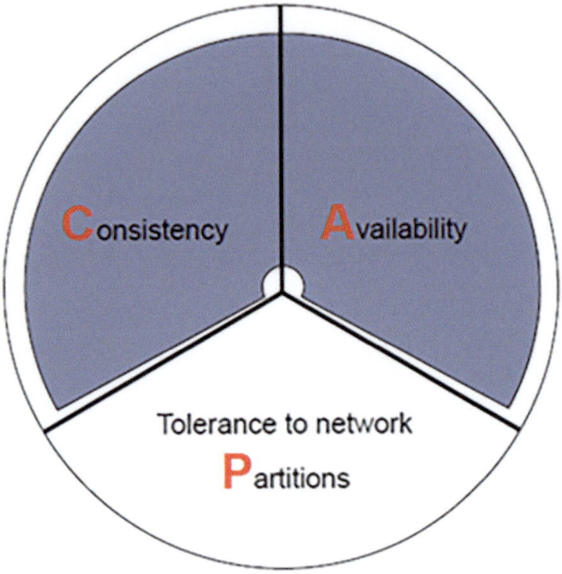

Fig. B.3 CA: Forfeit Partition

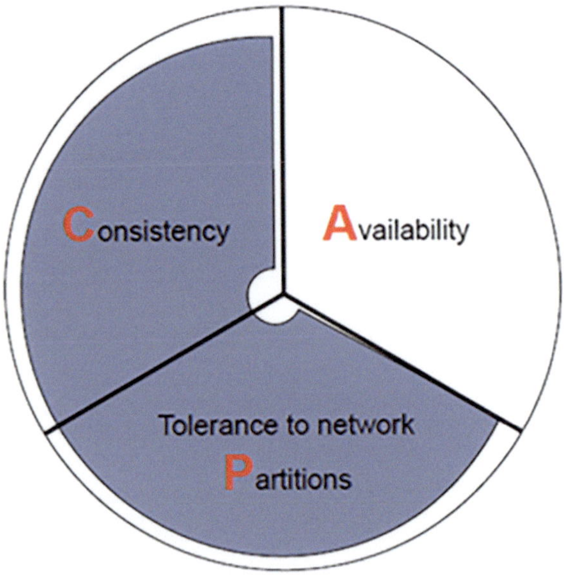

Fig. B.4 CP: Best effort availability

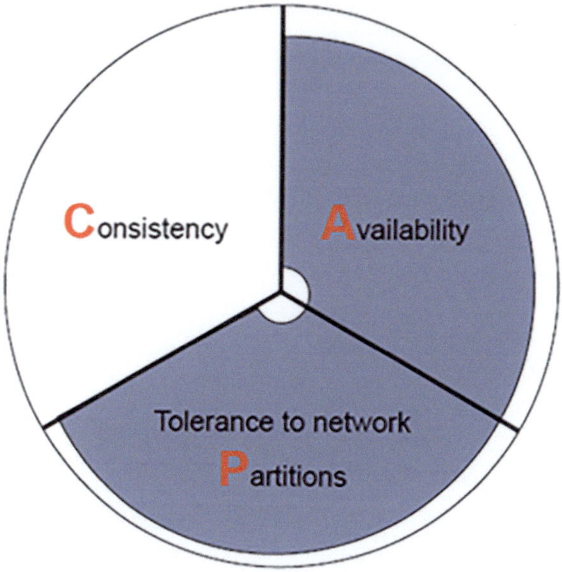

Fig. B.5 AP: Best effort availability

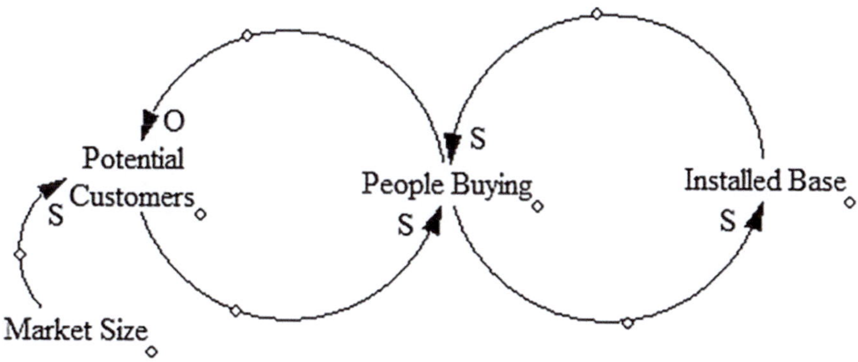

Fig. B.6 Casual loop diagram: Customers

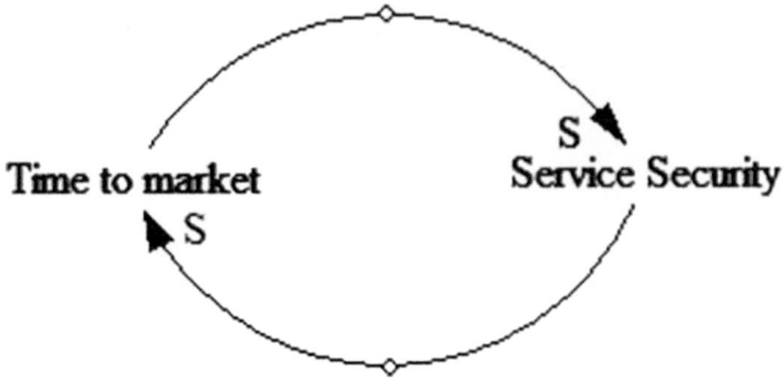

Fig. B.7 Casual loop diagram: Security

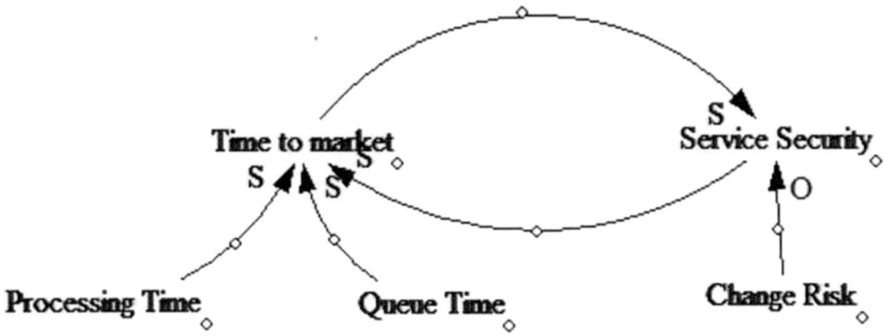

Fig. B.8 Casual loop diagram: Change risks

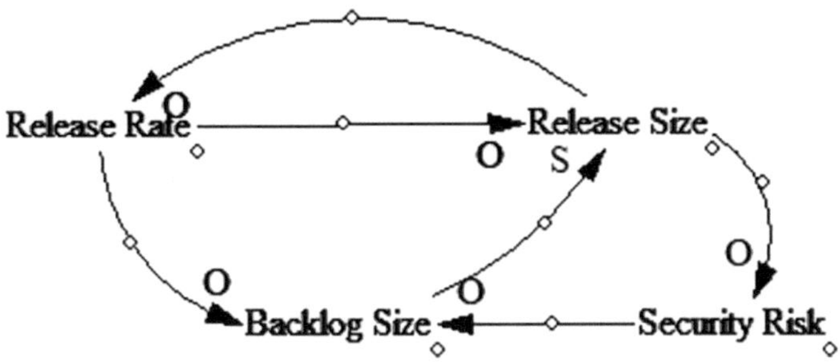

Fig. B.9 Casual loop diagram: Security risks

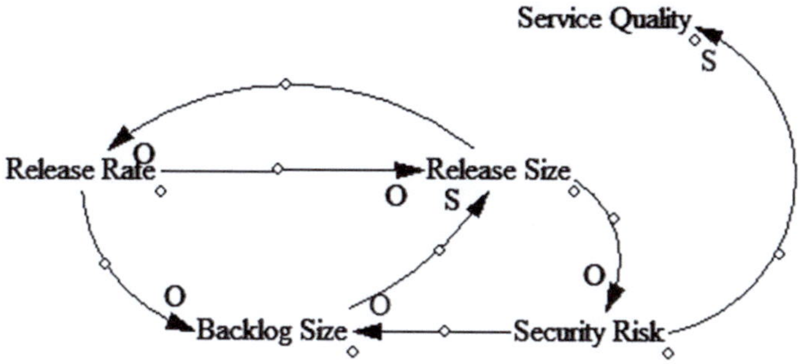

Fig. B.10 Casual loop diagram: Process time

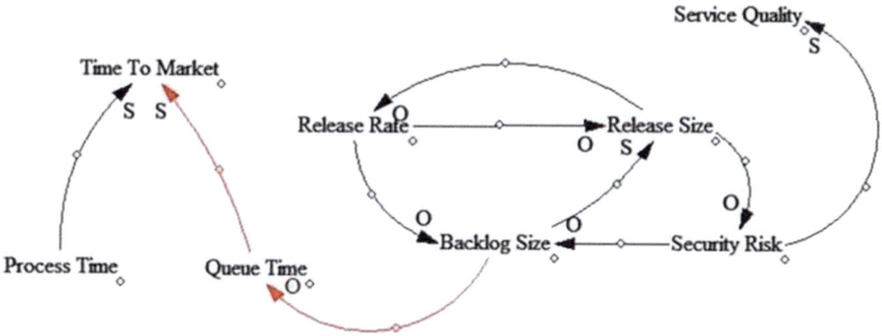

Fig. B.11 Casual loop diagram: Release rate

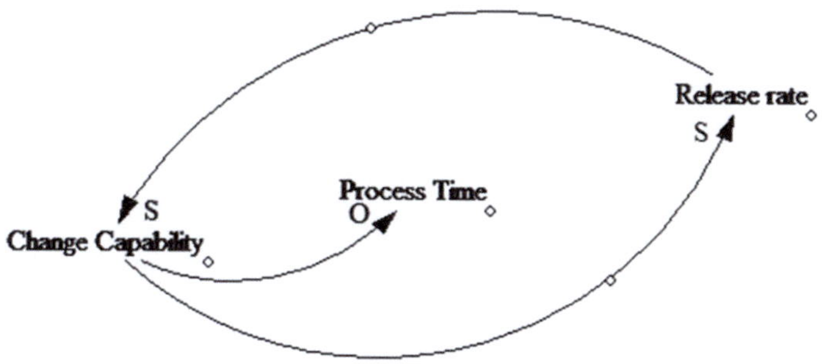

Fig. B.12 Casual loop diagram: Change and release rate

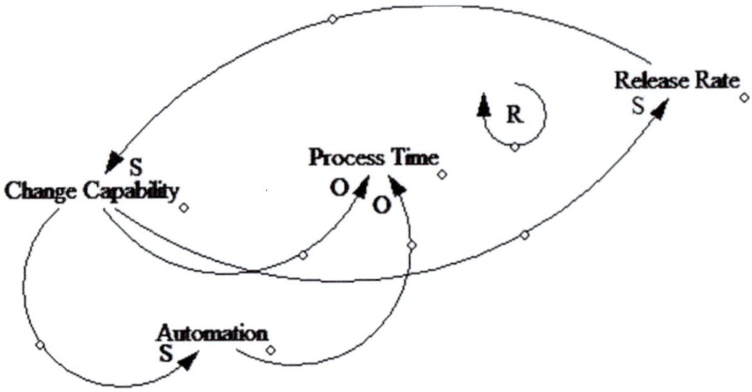

Fig. B.13 Casual loop diagram: Change management

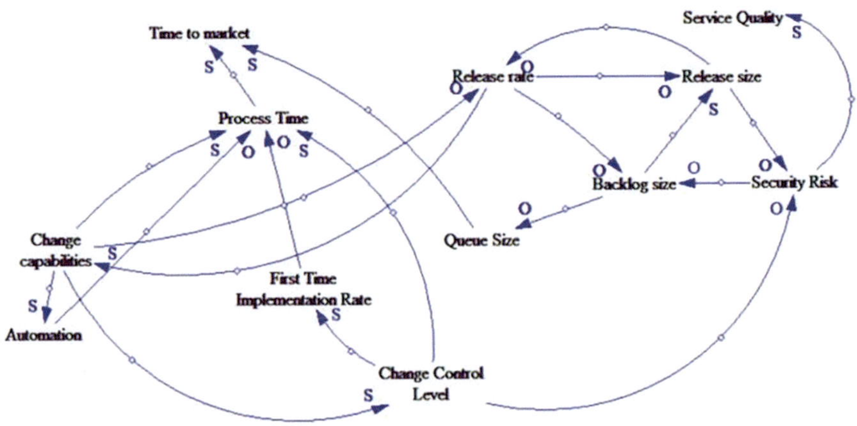

Fig. B.14 Casual loop diagram: time to market and service quality

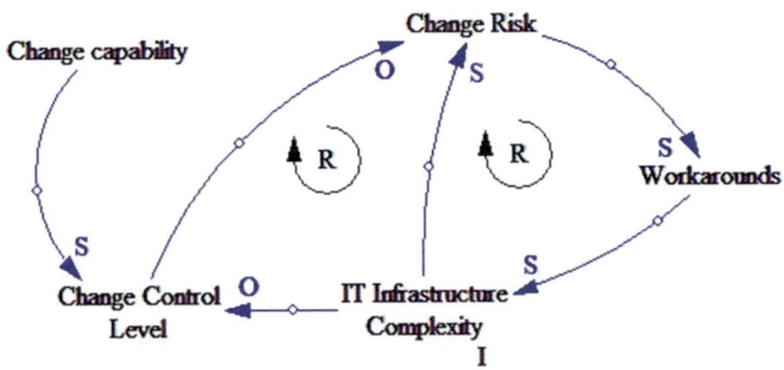

Fig. B.15 Casual loop diagram: Workarounds and Risks

Appendix C
Processes Re-Engineering for Hydraulic Fracturing

See Figs. C.1 and C.2.

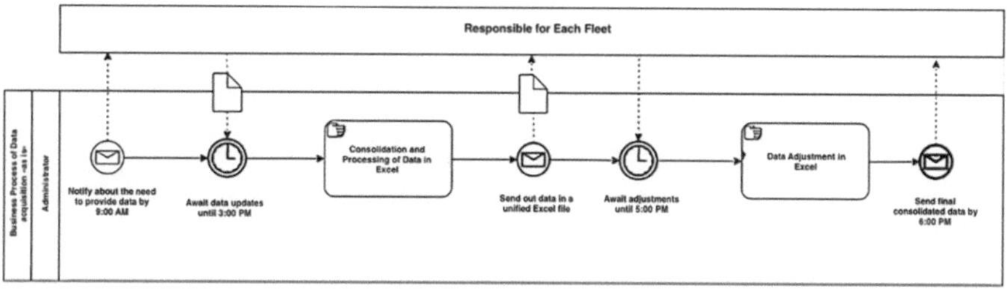

Fig. C.1 Current business process of data acquisition

S. V. Zykov, *IT Crisisology Patterns and Practices*, Smart Innovation, Systems and Technologies 449, https://doi.org/10.1007/978-981-95-1260-7

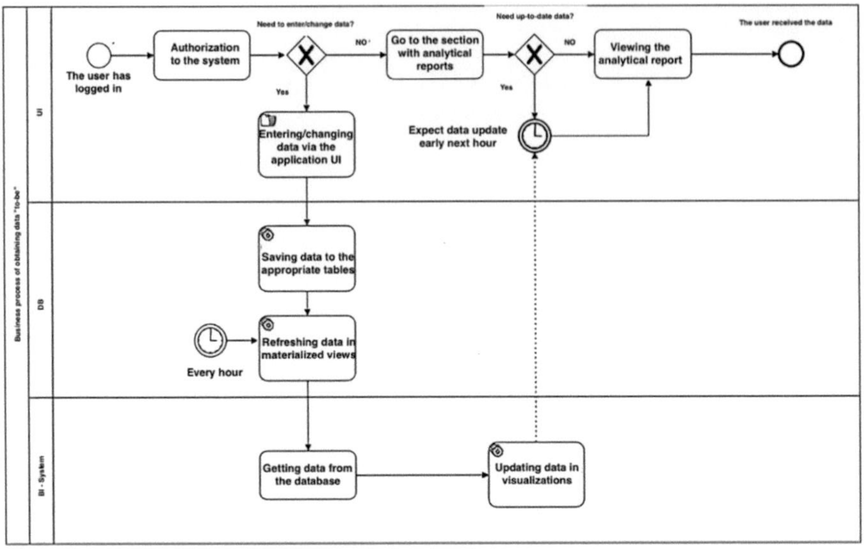

Fig. C.2 Target business process of data acquisition

Appendix D
Intelligent Portfolio Management for Financial Group

See Figs. D.1, D.2, D.3, D.4, D.5 and D.6.

S. V. Zykov, *IT Crisisology Patterns and Practices*, Smart Innovation, Systems
and Technologies 449, https://doi.org/10.1007/978-981-95-1260-7

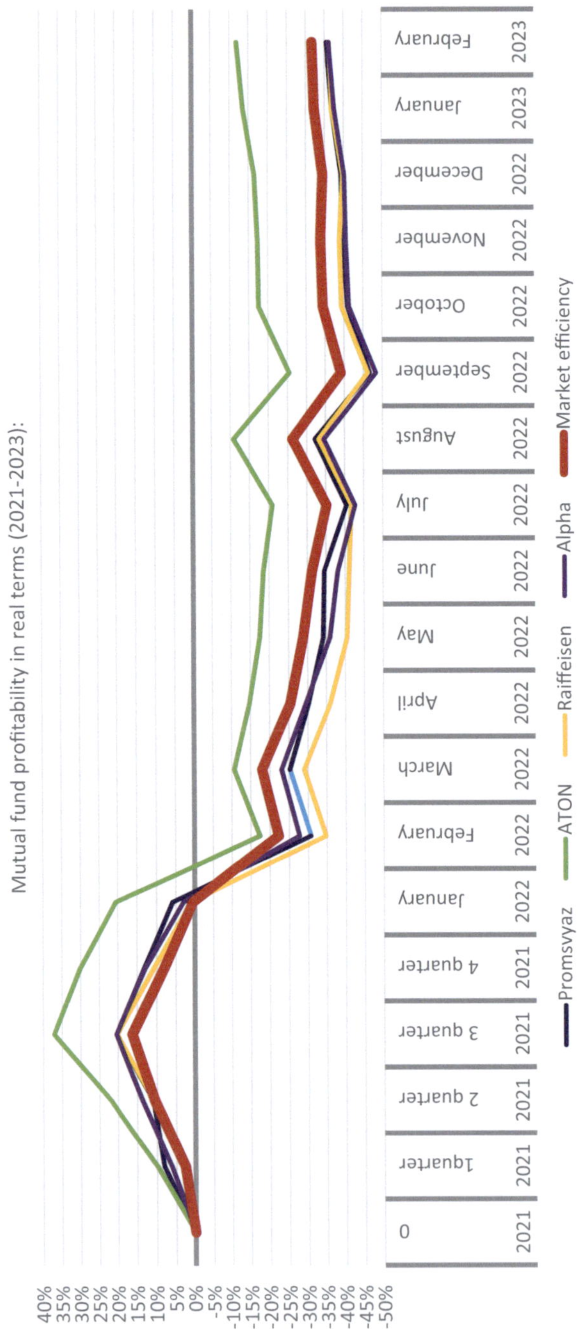

Fig. D.1 Mutual fund profitability in real terms (2021–2023)

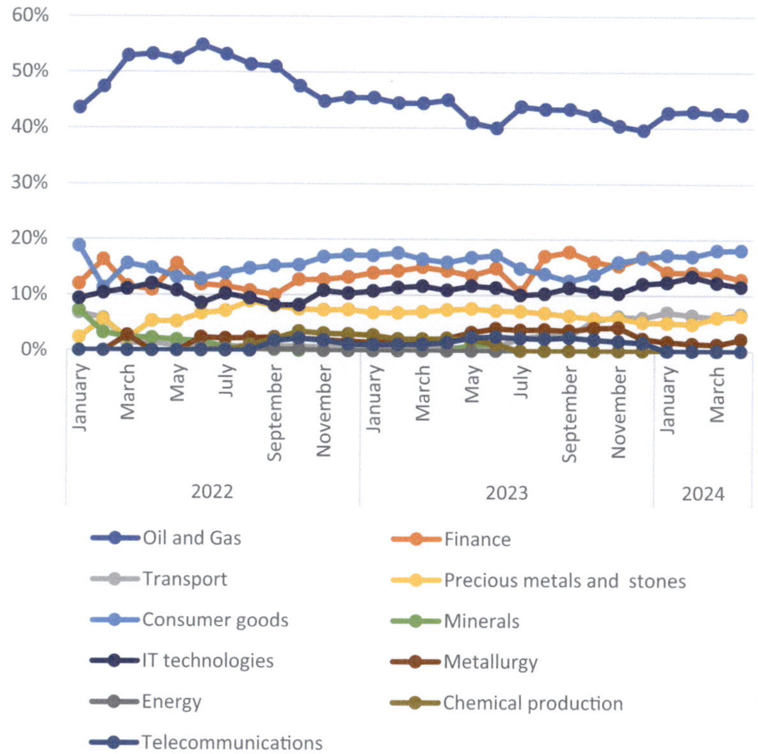

Fig. D.2 Portfolio structure in % by industry

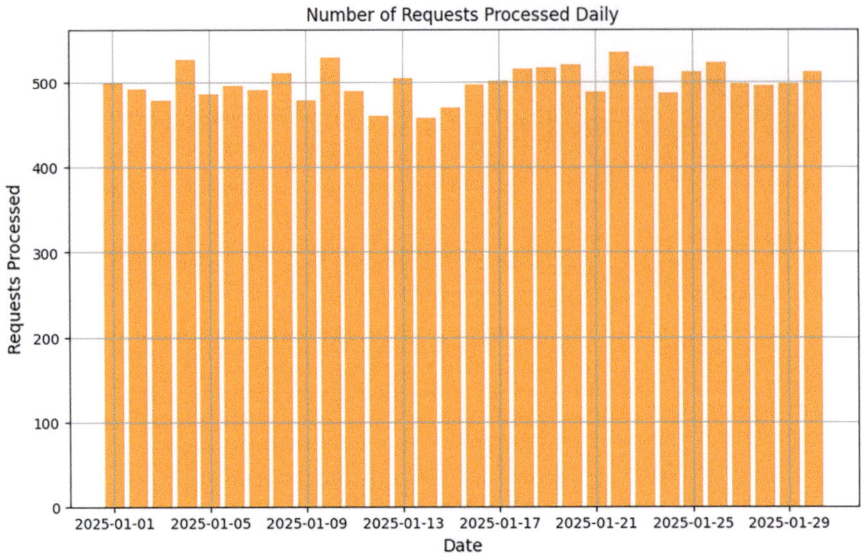

Fig. D.3 Number of requests processed daily

Fig. D.4 Distribution of user satisfaction scores

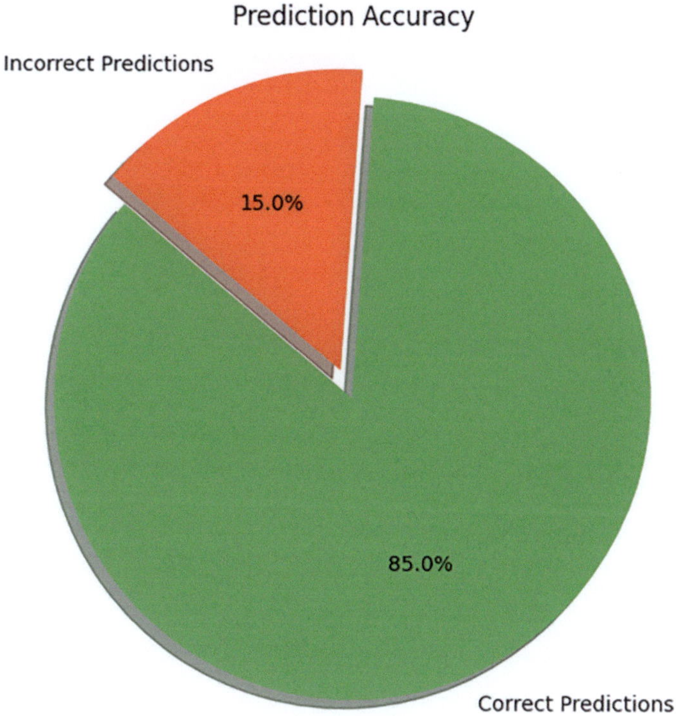

Fig. D.5 Prediction accuracy

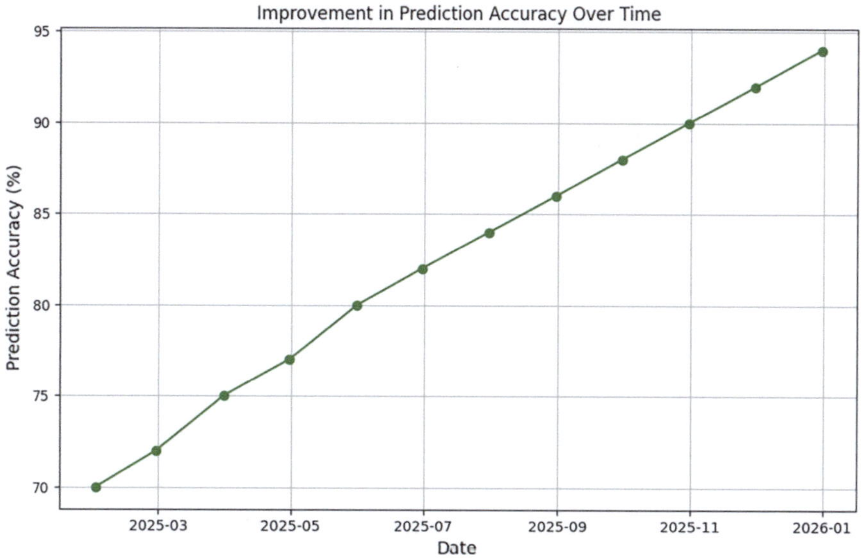

Fig. D.6 Improvement in prediction accuracy over time

Appendix E
IT Process Improvement for Service Company

See Figs. E.1, E.2, E.3, E.4, E.5 and E6.

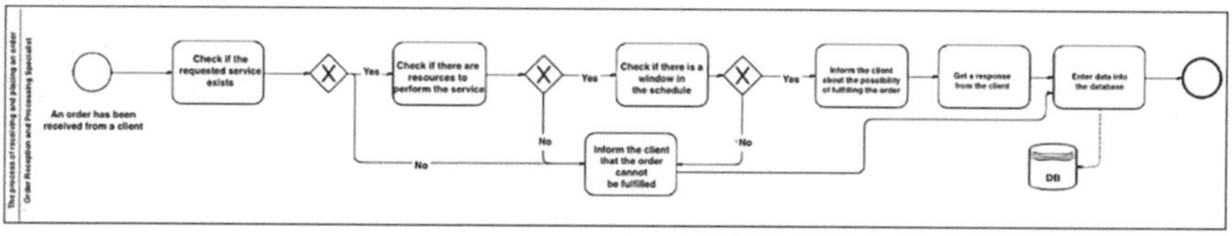

Fig. E.1 Process model of receiving and placing an order

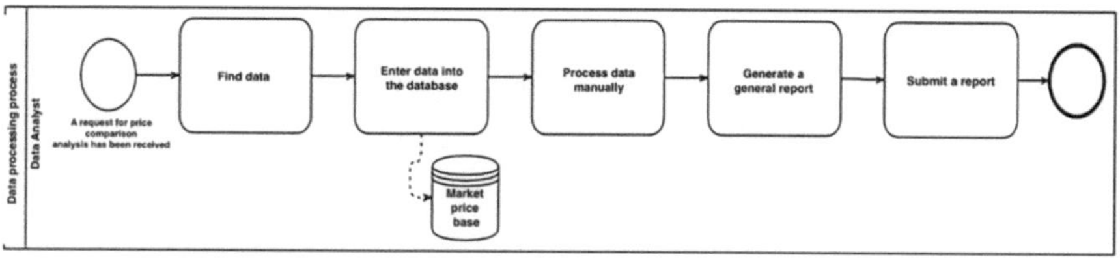

Fig. E.2 Process model of market price data processing

S. V. Zykov, *IT Crisisology Patterns and Practices*, Smart Innovation, Systems and Technologies 449, https://doi.org/10.1007/978-981-95-1260-7

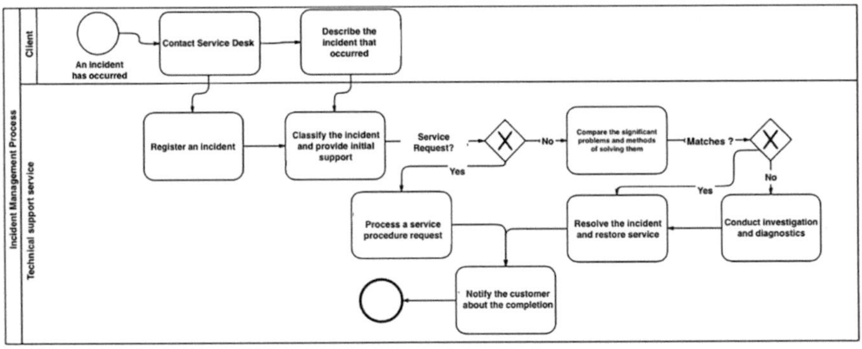

Fig. E.3 Process model of incident management

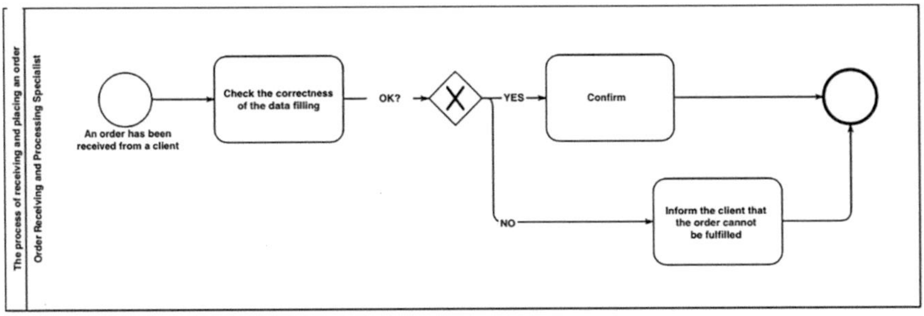

Fig. E.4 Optimized process model of receiving and placing an order

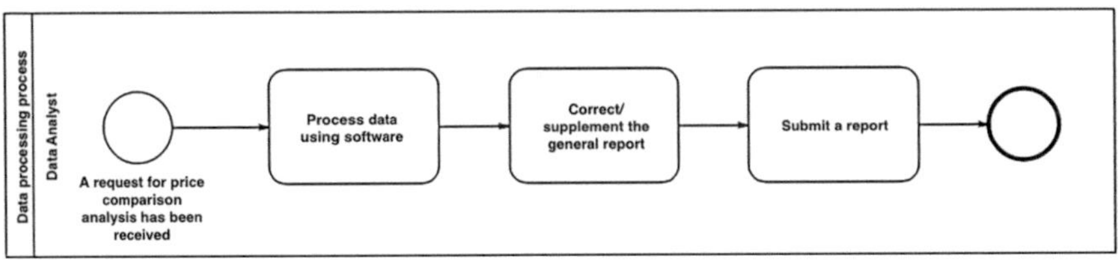

Fig. E.5 Optimized process model of data processing

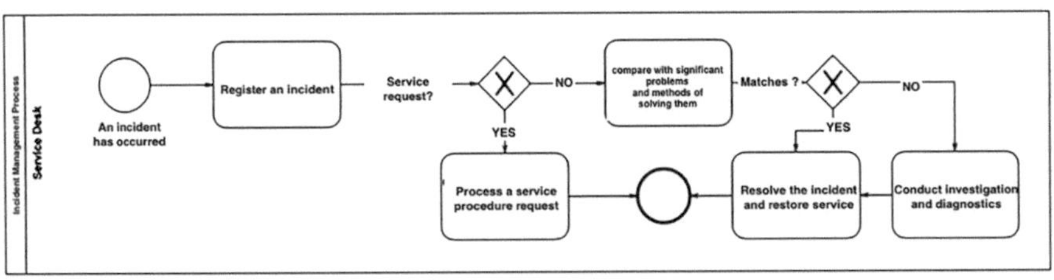

Fig. E.6 Streamlined process model of incident management

Appendix F
Data Integration Management

See Table F.1 and Figs. F.1, F.2, F.3, F.4, F.5, F.6, F.7, F.8, F.9, F.10, F.11, F.12, F.13 and F.14.

Table F.1 Comparison of competing tools

		Amundsen	DataHub	Metacat
Open source		+	+	+
Sources	Postgres	+	+	+
	MySQL	+	+	+
	HDFS	+	+	+
	Kafka	+	+	–
	CSV-files	+	–	–
	Oracle	+	–	–
	Snowflake	+	+	+
	Elastic	+	+	+
	Clickhouse	–	–	–
Pull/Push		Pull	Push and Pull	Push
API	Rest JSON	–	–	–
	Rest Thrift	–	–	+
User defined metadata	Connections for RDS	–	–	+
	Free form	+	+	+
Schema change	Auditing	–	+	+
	Notification	–	+	+

S. V. Zykov, *IT Crisisology Patterns and Practices*, Smart Innovation, Systems and Technologies 449, https://doi.org/10.1007/978-981-95-1260-7

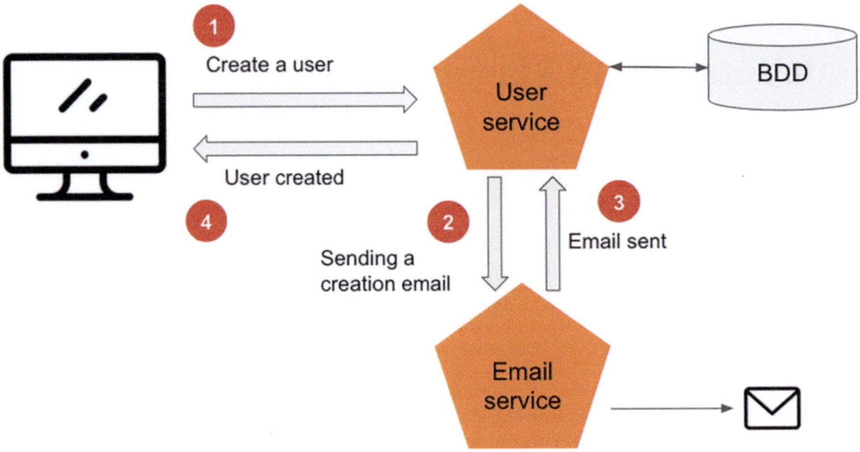

Fig. F.1 REST calls between the services

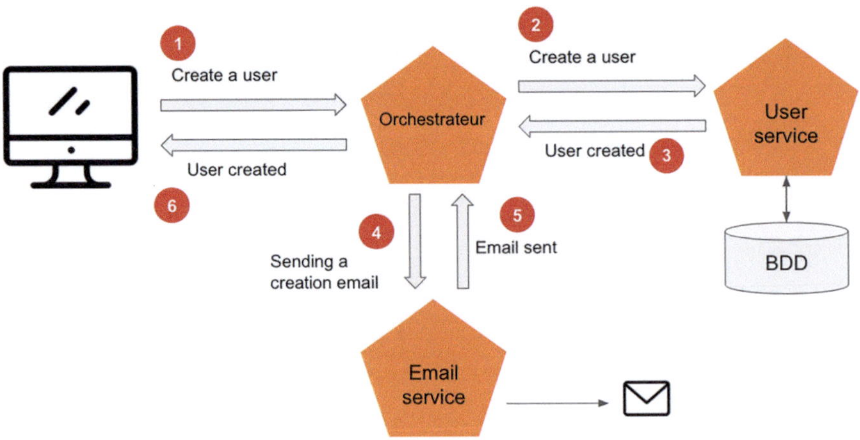

Fig. F.2 Orchestrator architecture

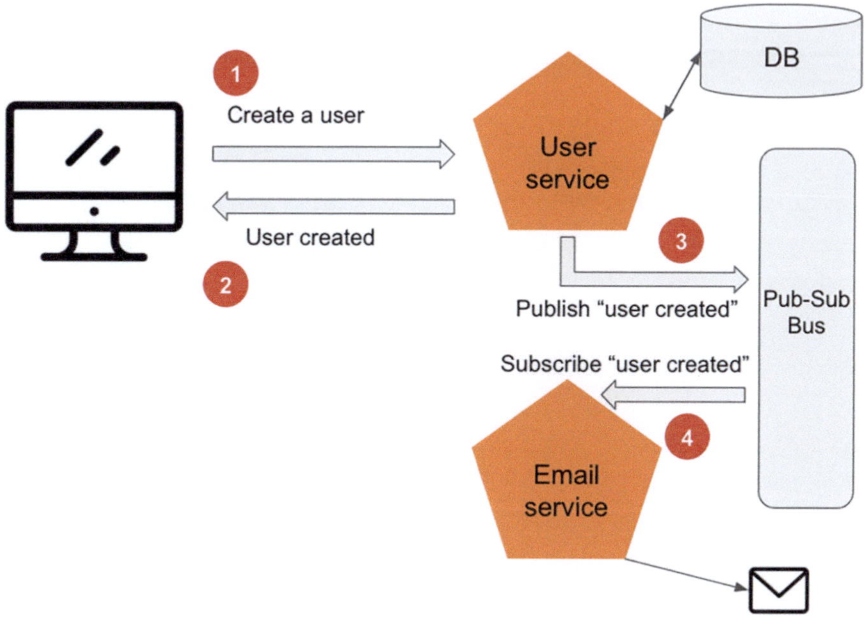

Fig. F.3 Publish–subscribe (Pub-sub) architecture

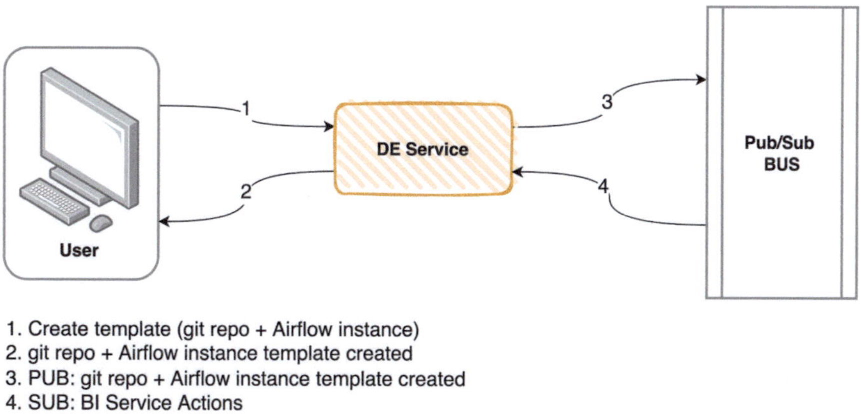

Fig. F.4 Data Engineering (DE)-based microservice interaction

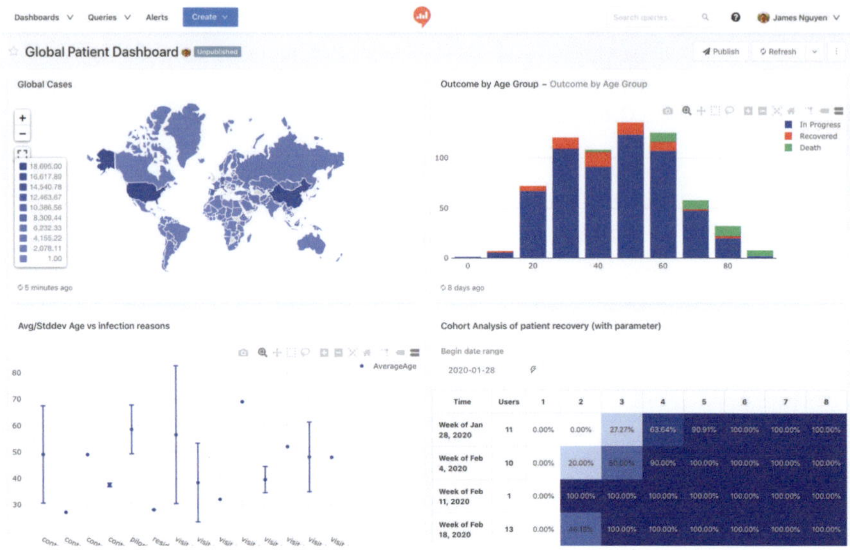

Fig. F.5 Redash example visualization dashboard

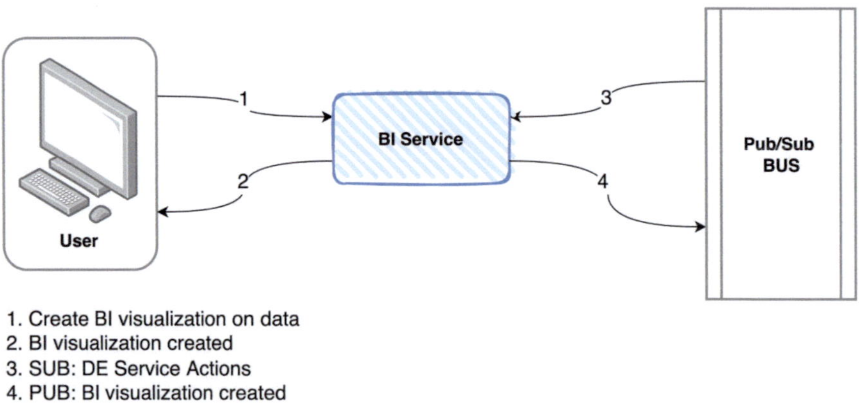

Fig. F.6 Business Intelligence (BI)-based microservice interaction

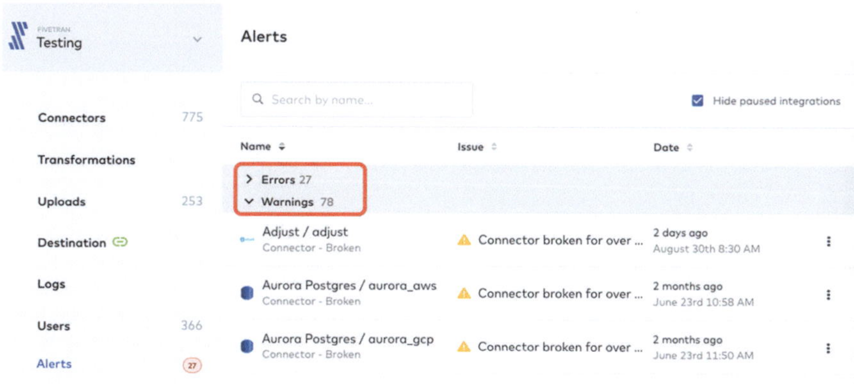

Fig. F.7 Fivetran: alerting features

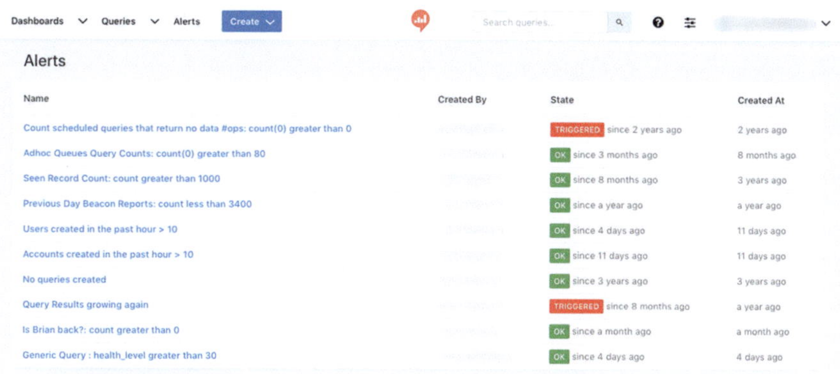

Fig. F.8 Redash: alerting features

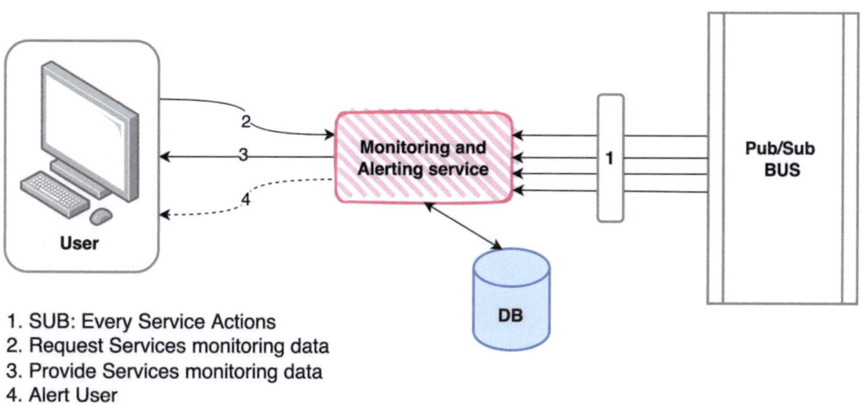

Fig. F.9 Monitoring and alerting microservice

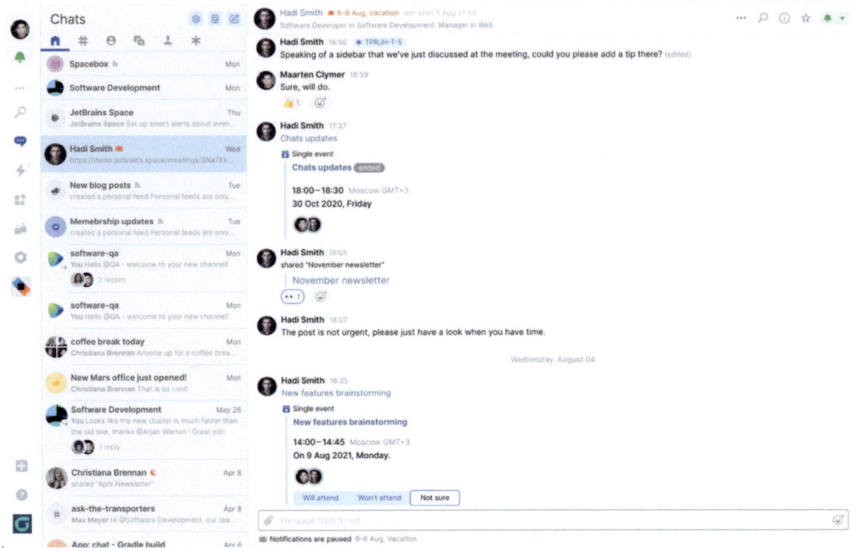

Fig. F.10 Chats in Space product by JetBrains

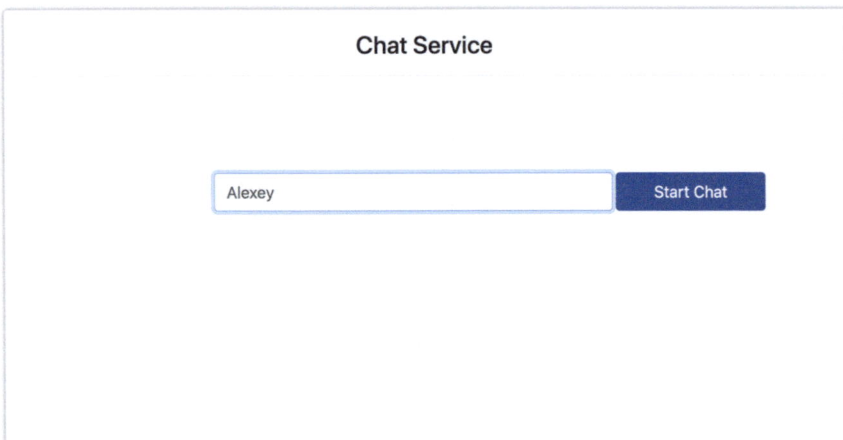

Fig. F.11 Logging interface

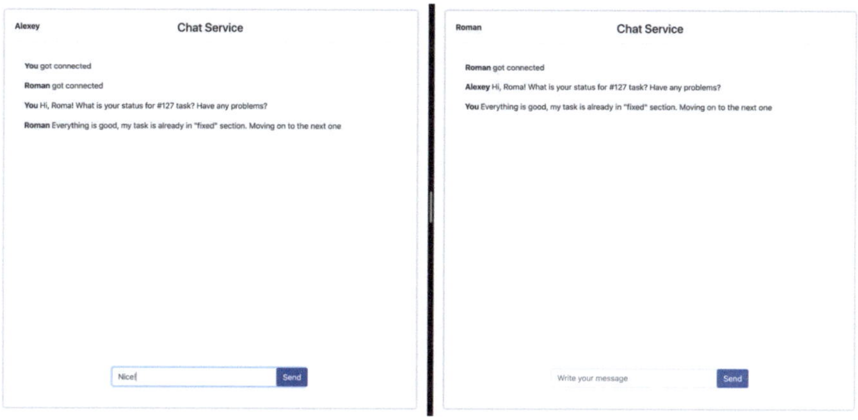

Fig. F.12 Messaging interface

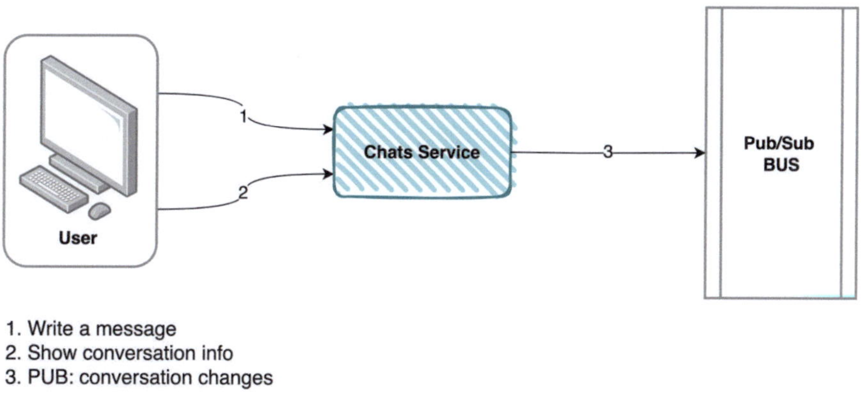

Fig. F.13 Chat microservice: messaging sequence

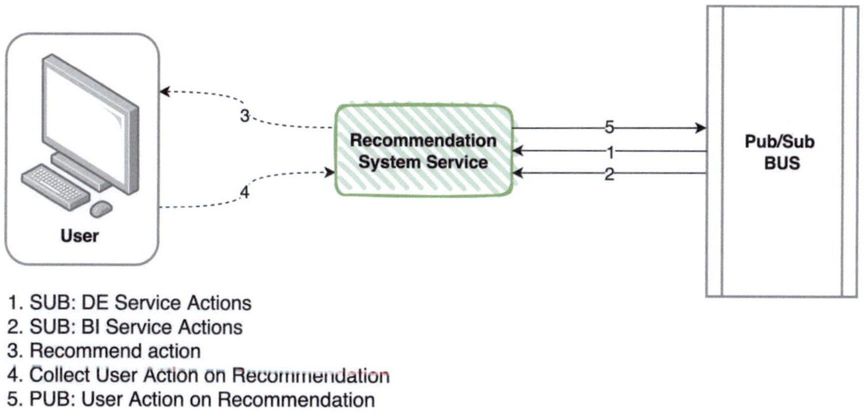

Fig. F.14 Recommendation microservice: sequence diagram

Appendix G
Patterns and "Antipasti"

See Tables G.1, G.2, G.3, G.4 and G.5.

Table G.1 System requirements

Stakeholder	Requirement	Priority
Logistics department	Real-time monitoring	High
IT	Easy integration with external services	Average
Management	System scalability	High
Analytics	Demand forecasting using ML	Low

Table G.2 SWOT analysis

Strengths	Weaknesses	Opportunities	Threats
IoT enables tracking of goods	Additional sensors required	Increase supply chain visibility	Equipment cost
AI improves demand forecasts	Complexity of implementation	Cost reduction	Algorithmic errors can affect the results

S. V. Zykov, *IT Crisisology Patterns and Practices*, Smart Innovation, Systems and Technologies 449, https://doi.org/10.1007/978-981-95-1260-7

Table G.3 Selection of technologies

Component	Technology	Justification
Microservice Orchestration	Kubernetes	Scalability and stability
Containerization	Docker	Easy deployment
Transactional data	PostgreSQL	Robust relational database
Unstructured data	MongoDB	Data flexibility
API Gateway	Kong	Centralized API management

Table G.4 Sprint plan

Sprint#	Sprint goal(s)	Time	Comments
1	Inventory management module(s) development	2 weeks	Includes database integration
2	Transport module(s) development	2 weeks	Communication with external vendor APIs
3	System integration and testing	1 week	Automated testing

Table G.5 Testing

Testing type	Description	Testing tool
Unit testing	Verify and validate individual microservices	JUnit
Integration testing	Verify and validate microservice interoperability	Postman, Newman
Load testing	Verify and validate system performance under high load	JMeter

Index

S. V. Zykov, *IT Crisisology Patterns and Practices*, Smart Innovation, Systems and Technologies 449, https://doi.org/10.1007/978-981-95-1260-7

FSC
www.fsc.org
MIX
Papier aus verantwortungsvollen Quellen
Paper from responsible sources
FSC® C105338

If you have any concerns about our products,
you can contact us on
ProductSafety@springernature.com

In case Publisher is established outside the EU,
the EU authorized representative is:
Springer Nature Customer Service Center GmbH
Europaplatz 3, 69115 Heidelberg, Germany

Printed by Libri Plureos GmbH
in Hamburg, Germany